Home Plumbing Illustrated

Home Plumbing Illustrated

R. Dodge Woodson

TAB BOOKS
Blue Ridge Summit, PA

BOCA National Plumbing Code is the registered trademark of the Building Officials and Code Administrators International, Inc.

FIRST EDITION
FIRST PRINTING

© 1993 by **R. Dodge Woodson**.
Published by TAB Books.
TAB Books is a division of McGraw-Hill, Inc.

Library of Congress Cataloging-in-Publication Data

Woodson, R. Dodge (Roger Dodge), 1955 –
   Home plumbing illustrated / by R. Dodge Woodson.
     p.   cm.
   Includes index.
   ISBN 0-8306-3986-1 (p)
    1. Plumbing—Amateurs' manuals.   I. Title.
TH6124.W67 1992
696'.1—dc20   92-15754
                                     CIP

TAB Books offers software for sale. For information and a catalog, please contact TAB Software Department, Blue Ridge Summit, PA 17294-0850.

Acquisitions Editor: Kim Tabor
Editor: Barbara B. Minich
Director of Production: Katherine G. Brown
Book Design: Jaclyn J. Boone
Cover Design: Graphics Plus, Hanover, Pa.
                                                              AV1

This book is dedicated to my wife and daughter. Afton, my daughter, while not yet four, has a strong sense for the importance of this and all of my books. Her cooperation and presence provides constant support for my writing. My wife, Kimberley, has always been there for me. Her consideration makes a project like this not only possible, but enjoyable.

Contents

Part II Planning, installation, and inspections of new plumbing jobs

16 Removing existing plumbing and fixtures — 200

17 Remodeling installation variations — 214

Part IV Troubleshooting pointers

18 Common malfunctions with DWV systems — 221

Acknowledgments

First, I would like to acknowledge and thank my parents, Woody and Maralou, for their role in my life. Fran Pagurko is thanked for his cooperation in modeling for many of the photos in this book.

The following is a list of companies considerate enough to supply quality art to illustrate much of the text:

American Standard, Inc.
P.O. Box 6820
Piscataway, NJ 08855-6820

Fernco
300 S. Dayton St.
Davison, MI 48423

General Wire Spring Co.
1101 Thompson Ave.
McKees Rocks, PA 15136

Goulds Pumps, Inc.
P.O. Box 330
Seneca Falls, NY 13148

Hellenbrand Water Conditioners
404 Moravian Valley Rd.
Waunakee, WI 53597

Moen Inc.
377 Woodland Ave.
Elyria, OH 22036-2111

Ridge Tool Co.
400 Clark St.
Elyria, OH 22036

Vanguard Plastics
831 N. Vanguard St.
McPherson, KS 67460

Universal Rundle
303 North St.
New Castle, PA 16103

Introduction

Welcome to one of the most comprehensive plumbing books available. If you have ever wished for an in-depth guide to plumbing that you could understand, this is it. This book is written so that the average person can perform professional plumbing techniques. The book was written by a master plumber, with 18 years of field experience, who has taught apprenticeship and plumbing code classes in state vocational schools. Woodson has also written many how-to books and articles. With his broad knowledge and easy-to-understand instructions, Woodson guides you through all the steps needed to handle most plumbing jobs.

This book is not written in a cold, cryptic style. Instead, the author talks to you and guides you along the path of professional plumbing applications. You will learn the necessary steps to plumb a new house. Even if you only want to install a new well pump, the procedure is covered in detail. Other chapters deal with troubleshooting and correcting deficiencies in your plumbing system.

Regardless of your level of experience, having this book is like having a master plumber working beside you. If you are in the middle of a project and become confused, you can turn to the section of the book that covers your problem area. With its many subheadings and thorough index, locating the information you need is fast and simple. Whatever your plumbing job, this book will make it easier.

If technical data is what you are after, it's here. Charts and tables help you design and size a complete plumbing system. While the information is technical, it is delivered in an easy-reading style. You will experience what it is like to be on the job when looking at the numerous photographs and illustrations. These visual aids will make understanding the text even easier.

Part I

Tools and approved plumbing materials

1

The toolbox

PLUMBING IS A TRADE that requires a few specialized tools, as well as many tools found in average households. As we go through this chapter, you will learn about the tools most frequently found in a plumber's toolbox. I will explain how the tools will be used in the plumbing trade. When we come across a tool that has no effective substitute, I will bring it to your attention. Most specialized plumbing tools can be rented from local rental centers.

Hand tools

Hand tools are the tools most often used by plumbers. These tools include items like hacksaws and hammers. What follows is a listing and description of all the most common hand tools carried by working plumbers.

Hacksaws

Hacksaws (Fig. 1-1), which are used to cut plastic pipe, metal pipe, bolts, and just about anything else that a plumber needs to cut, see many hours of use by professional plumbers. For part-timers, any common hacksaw will be adequate, but for heavy-duty use, plumbers require rugged saws. Adjustable hacksaws, used by most plumbers, take various sizes of blades, and the blades can be mounted in different ways.

The most popular adjustable hacksaws allow the blade to be installed to cut straight or at an angle. Angle mounting the blade makes it simpler to cut closet bolts and similar items. The saw can be held in a normal, upright position while cutting across the base of a toilet. Installing the blade to protrude from the end of the saw allows cutting in tight spaces. The only clearance required when the blade is mounted in this fashion is clearance for the blade.

When you select blades for your hacksaw, consider the types of material you will be cutting. If you are cutting metal, a blade with a high tooth count will work best. Cutting plastic pipe is easier and faster with a blade that has few teeth.

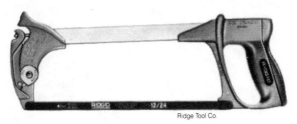

Fig. 1-1. Hacksaw.

Roller-type cutters

Roller-type cutters are used primarily for cutting copper and steel pipe, but they can be used to cut plastic pipe. Roller cutters come in different sizes to accommodate various types and sizes of pipes (Fig. 1-2). The roller cutter is the best tool to use to cut copper pipe and tubing. Copper tubing and pipe can be cut with a hacksaw, but this leaves ragged edges, making cutting more difficult. Roller cutters cut copper cleanly and quickly.

Fig. 1-2. Copper tubing cutter.

Roller cutters for steel pipe are larger than those used for copper (Fig. 1-3). They have a long handle for leverage and require a fairly large turning radius around the pipe. If you are cutting metal pipe already installed, there will rarely be room to use a roller cutter. In this case, you will probably use a hacksaw or a reciprocating saw with metal-cutting blades.

Fig. 1-3. Steel pipe cutter.

Plastic pipe can be cut with roller cutters but most plumbers use a saw. There are even roller cutters designed to cut cast-iron pipe but, like steel cutters, the roller cutters for cast-iron pipe require extensive turning room around the pipe. When it comes to cutting cast-iron, there are other tools designed for such situations.

Miniature roller cutters

Miniature roller cutters can save the day when you work with copper in tight spaces (Fig. 1-4). These mini-cutters are ideal for cutting copper in places where normal cutters cannot be rotated around the pipe. If you buy a pair of mini-cutters, buy good ones. There is not much opportunity to get leverage on these tiny cutters, so you need the best cutters you can get.

Fig. 1-4. Mini-cutter for copper tubing.

Internal pipe cutters

Internal pipe cutters are not needed often, but when they are, they are indispensable (Fig. 1-5). An example would be a pipe in a concrete floor that is too tall for a water closet. If you need to cut the pipe close to or below the floor, internal cutters are the best tool for the job. The cutter fits into the top of the pipe and is adjusted to apply pressure between the cutting wheel and the pipe. As the tool is turned, the handle is tightened to maintain cutting pressure. After several turns, the pipe cuts cleanly.

Fig. 1-5. Internal tubing cutters.
Ridge Tool Co.

Cast-iron pipe cutters

There are several ways to cut cast-iron pipe, but one method is easier than all the rest. To cut cast-iron pipe easily, you will need a rachet soil pipe cutter (Fig. 1-6). If you don't have a soil pipe cutter, you will be forced to use more primitive methods to cut cast-iron pipe. Rachet cutters are not inexpensive, but they can be rented from most rental centers.

Fig. 1-6. Ratchet-style soil pipe cutter.

You don't need a lot of space to work with a rachet cutter. A chain with cutting wheels is wrapped around the pipe and attached in the notches on the tool. A wheel is turned to apply pressure to the pipe with the cutting wheels. The handle is then cranked up and down to cut the pipe, cutting the pipe in a matter of moments.

If you don't use a rachet cutter, you will have to cut the pipe with other tools. The next best tool is a snap cutter (Fig. 1-7). Snap cutters use the same type of chain and cutting wheel assembly but require more space and strength to work. Snap cutters have two handles that spread out like open scissors. When the two handles are compressed, the pipe is cut.

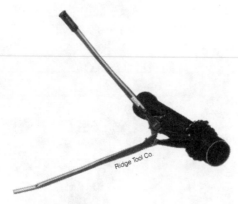

Fig. 1-7. Snap-style soil pipe cutter.

Special blades are available for reciprocating saws that cut cast-iron pipe. With a good blade, you can expect a single cut to take up to 20 minutes. Some old-timers still know how to cut cast-iron with a hammer and a chisel, but there are not many people around with this skill. If you have to cut cast-iron pipe, rent a rachet cutter.

Flaring tools

You will not usually need a flaring tool for residential plumbing. The exception is plumbing in mobile homes. Many trailers contain flare joints. If you need to flare tubing, a flaring tool is essential (Fig. 1-8).

Fig. 1-8. Flaring tool.

Pipe wrenches

Pipe wrenches are not needed as much for modern plumbing installations as they were in past years. Aluminum pipe wrenches are a little more expensive than standard ones, but they are much lighter and easier to use. If you want to stock your toolbox with common pipe wrenches, you will want four different sizes. If you buy 10-inch, 14-inch, 18-inch, and 24-inch pipe wrenches, you will be prepared for most any circumstance where a pipe wrench is required.

Chain wrenches

For new-construction plumbing, you will not need a chain wrench. If you are doing remodeling work, you might find chain wrenches useful (Fig. 1-9). They can sometimes be used in place of a pipe wrench when space is at a premium.

Fig. 1-9. Chain wrench.

Ridge Tool Co.

Strap wrenches

Strap wrenches fall into about the same category as chain wrenches. You will not need a strap wrench for new-construction plumbing, but you might like to have one when working with existing plumbing.

Basin wrenches

Basin wrenches are almost a necessity when it is time to set fixtures (Fig. 1-10) and are invaluable when connecting the supply nuts to faucets from beneath a sink. If you will be doing extensive plumbing, invest in a good, telescoping basin wrench. If you will only be using the tool for one job, you can buy a less expensive wrench that is not adjustable. The adjustable style affords you more comfort and flexibility in various plumbing situations.

Fig. 1-10. Basin wrenches.

Offset hex wrenches

Offset hex wrenches are a special tool that most plumbers don't even carry (Fig. 1-11). They are used to turn large nuts, like those used to mount basket strainers to a kitchen sink. Most plumbers use large tongue-and-groove pliers instead of an offset hex wrench. If you don't mind investing in tools, these wrenches are helpful for turning large nuts.

Fig. 1-11. Offset hex wrench.

Adjustable wrenches

Adjustable wrenches, used to set fixtures, repair work, and occasionally to rough-in plumbing, are used frequently in the plumbing trade. Stock your toolbox with assorted sizes of these regularly used tools.

Metal-cutting snips

Metal-cutting snips are used to cut metal strapping to hang pipes and for other metal cutting jobs. There are two types of snips to consider purchasing (Fig. 1-12). The first type is designed to make straight cuts. For most plumbing work, these snips are all you need. The other types are offset snips. They are made to cut left or right. Sheet metal workers use both in their work, but plumbers can get by with the type that cuts straight.

Fig. 1-12. Metal-cutting snips. Ridge Tool Co.

Utility pliers

Normal utility pliers are not a major asset to a plumber.

Needle-nose pliers

Needle-nose pliers are essential in plumbing work (Fig. 1-13). Many jobs require their thin design to reach awkward objects. Most plumbers carry at least two sizes of needle-nose pliers.

Fig. 1-13. Needle-nose pliers.
Ridge Tool Co.

Tongue-and-groove pliers

Tongue-and-groove pliers, the most often-used tool in the plumber's toolbox, are a plumber's workhorse (Fig. 1-14). These offset pliers are used for everything from opening cans of glue to tightening slip nuts. You should have at least two pairs, both large enough to use on a 2-inch nut. If you can afford additional sets, buy a large pair and a small pair.

Fig. 1-14. Tongue-and-groove
pliers. Ridge Tool Co.

Spring tubing benders

Most professionals don't use spring benders, but for inexperienced people, spring benders make bending supply tubes easier (Fig. 1-15). If you are not accustomed to bending sensitive, thin-wall tubing, spring benders can reduce the risk of damaged tubing. The supply tube is placed inside the spring bender and then bent with even pressure to avoid crimping the tube.

Ridge Tool Co.

Fig. 1-15. Spring tubing bender.

Tape measures

Every plumber has a measuring device. Most plumbers use a retractable, metal measuring tape. The average tape measure is 25 feet long with a one-inch-wide blade. The wider blade allows the tape to be extended for a longer distance before it bends. Because many measurements must be made from a ladder, the longer the blade, the less you will have to climb up and down the ladder.

In addition to a retractable tape measure, a long, rolled tape measure, used to lay out walls, measure sewers, and for any other application where the measurement is a long one, comes in handy for some plumbing jobs. These tapes are usually either 50 or 100 feet long.

Grade levels

Grade levels can be considered a specialty tool used by professionals to speed up the piping process and ensure accuracy and consistency in the pipe's grade (Fig. 1-16). They are equipped with an adjustable, threaded rod, that allows the level to be used to set the grade on drainage pipe.

Fig. 1-16. Grade level.

How do grade levels work? Assume you have a 2-foot grade level and want to install your pipes with a grade of 1/4 inch per foot. Turn the threaded adjustment stud down until it protrudes 1/2 inch below the surface of the level. When you install the pipe, place the level on the pipe with the adjustment stud on the end running to the sewer. When the grade of the pipe is 1/4 inch to the foot, the bubble in the level will read level. Because the adjustment is 1/2 inch below the surface of the 2-foot level, the level reads level if the pipe drops 1/4 inch in each foot that it runs.

Do you need a grade level? Not unless you are planning to do a large volume of plumbing. For occasional jobs, you can tape a 1/2-inch-thick piece of wood to one end of a conventional, 2-foot level to obtain similar results.

Torpedo levels

Torpedo levels, used to check pipe grade, set fitting on grade, and to check fixtures for a level setting, are small, inexpensive, and very handy (Fig. 1-17). The most specialized use of a torpedo level is to set the grade of a fitting. If you are setting a fitting with a 45-degree angle, a torpedo level can tell you when the fitting is in the proper position. Because torpedo levels are short, they are not the most accurate level to use in pipe grading or fixture setting. These small levels are used frequently by working plumbers and should be in your toolbox.

Fig. 1-17. Torpedo level.

Ridge Tool Co.

Conventional levels

When it comes to conventional levels, most plumbers will carry a 2-foot and a 4-foot level. The 2-foot level is used for setting sinks, toilets, and other small- to moderate-size fixtures. The 4-foot level is used to set bathtubs and showers.

Line levels

Line levels, which hang on a piece of string, are used by plumbers to install underground plumbing (Fig. 1-18). When pipes must be kept below a concrete floor, a line level is indispensable. You can stretch a string across your work area to simulate the concrete floor.

Fig. 1-18. Line level.

Square-blade screwdrivers

Square-blade screwdrivers are used for numerous plumbing operations. Your toolbox should contain several screwdrivers in various sizes. The length of the shafts and the size of the blades should be diverse enough to meet any need.

Phillips screwdrivers

Phillips screwdrivers fall into the same category as square-blade screwdrivers. You should have an assortment of Phillips screwdrivers in your tool collection.

Convertible screwdrivers

Convertible screwdrivers are the ones most often found in the tool belt of most plumbers. These screwdrivers are equipped with interchangeable bits. Most convertible screwdrivers come with two sizes of flat-blade bits and two sizes of Phillips bits. Although these universal screwdrivers won't fill all of your needs, they will cover most of them.

Wood chisels

Wood chisels, used when installing pipe and fixtures, are required for some plumbing installations. You will not need superior chisels like those used by master carpenters, but an assortment of wood chisels will be a welcome addition to your toolbox.

Cold chisels

Cold chisels are sometimes needed to make holes in brick, block, or concrete. If you will be cutting holes through these types of materials, you will need a strong cold chisel. In place of a cold chisel, you might use a star drill or an impact drill.

Shovels

If you are installing underground plumbing, you will need a shovel. For most plumbing trenches, a round-point shovel is the best choice for the job.

Hammers

Hammers are constantly used in plumbing installations. Any type of carpenter's hammer will work, but I prefer a straight-claw, 20-ounce, framing hammer. In some cases, you might need a sledgehammer. The two sledgehammers most often used by plumbers are 8-pound and 2^1/$_2$-pound hammers.

Universal saws

Universal saws are made to cut plastic pipe (Fig. 1-19). Many plumbers use these saws, but I use a hacksaw. If I am cutting pipe larger than 4 inches thick, I use a regular handsaw. If you don't have a handsaw, universal saws are not too expensive and will get the job done.

Fig. 1-19. Universal saw.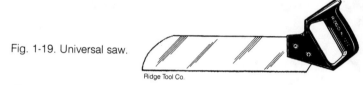

Ridge Tool Co.

Torches

If you will be soldering copper pipe, you'll need a torch. Torches can be fueled with propane or acetylene. For casual plumbing, hand-held torches that screw onto a propane bottle are sufficient. If you plan to get into heavier plumbing, invest in a professional torch kit. I prefer working with an acetylene torch, but many plumbers choose propane torches. Either of these torches will serve the purpose.

Ladders

You will need a solid stepladder, and possibly an extension ladder. Stepladders are used in many day-to-day activities. An extension ladder is usually only needed to install flashings on the vents penetrating the roof. Invest in good ladders; your safety is at risk with low-quality ladders.

Power tools

Plumbers do not require many power tools. A ^{1}/$_2$-inch, right-angle drill and a reciprocating saw are generally the only power tools needed to plumb a new house. The reciprocating saw is used to cut holes in the roof for vents, holes in the floor for closet flanges, and other miscellaneous cutting that might be required. A right-angle drill allows you to drill large holes in floor joists and walls where a straight drill could not operate.

Right-angle drills

While we are on the subject of right-angle drills, there are some safety concerns you should be aware of. Large, right-angle drills and bits used for plumbing can be very dangerous (Fig. 1-20). The drill develops a lot of torque and the bits can inflict nasty wounds. Before you drill a hole with these powerful machines, be sure there is nothing in the area to be drilled that could cause trouble. Nails and knots in the wood are common causes of drill-related injuries. Be sure all of your drill bits are sharp; dull bits increase the risk of injury.

Fig. 1-20. Right-angle drill.

Ridge Tool Co.

Never overextend your body when drilling. If the drill hits a nail or a knot, it can do severe damage to your body from the torque. Be especially careful if you are drilling through plywood. Once the bit's worm-driver is through the thin plywood, the bit can bite into the wood and jump back at you. I have seen many accidents caused by right-angle drills and their bits.

Reciprocating saws

Plumbers rely on reciprocating saws for many tasks (Fig. 1-21). These saws are ideal for cutting objects where space is limited. While it is possible to plumb a house without a reciprocating saw, there are many occasions when a reciprocating saw will save time and make difficult jobs easier.

Fig. 1-21. Reciprocating saw.

Ridge Tool Co.

Hand drills

Standard hand drills are used in some plumbing activities, but most drilling is done with a right-angle drill. If you are purchasing a hand drill, I recommend a 3/8-inch, variable-speed model. If you have a right-angle drill, there will be few occasions when you will need a hand drill.

Miscellaneous equipment

Some additional equipment you'll need to maximize the use of your tools are drill bits, saw blades, a striker, etc. Drill bits will be a necessity. There are many types of bits to choose from. Select bits that meet the requirements of your job. Saw blades will need to be replaced when they become dull, bent, or broken.

Many people light their torch with a match or cigarette lighter, but I prefer a striker. A striker intended for lighting torches provides some additional safety over matches and cigarette lighters. Your hand will be farther from the flame when it is ignited. This extra distance may help you avoid injury if the torch tip malfunctions.

As you get into the job, you may find you need additional tools: a plumb bob to align your holes; a chalk box to mark floor joists for drilling holes or spacing pipe. These miscellaneous items can be added to your toolbox as they are needed. While the tools described above may not be all the tools you will require, they represent most of the tools needed by a working plumber.

A final word on safety

Any job can present risks when proper safety rules are not observed. When you buy or rent a new tool, become familiar with its operation and safety guidelines before you put it into service. Safety glasses should be worn in many plumbing circumstances. These glasses protect your eyes from hot solder, wood splinters, and a number of other potentially injurious materials.

2

Approved materials

BEFORE YOU BEGIN your plumbing project, you must know what types of materials are approved for various jobs. A material that is approved for one purpose might be illegal for another. For several aspects of plumbing, you may choose from a wide variety of approved materials. This chapter covers the most commonly used materials and their applications.

Product markings

All materials used in an approved plumbing system must be marked to identify the manufacturer's name or logo. In addition to this marking, the quality of the material or other identification proving the acceptability of the material is required. The markings must be indelible. They can be embossed, stamped, or cast into the material.

For example, copper pipe is labeled with a colored stripe that identifies the type of copper. A red stripe indicates a type "M" copper. A blue stripe represents a type "L" copper, and yellow identifies the pipe as "DWV" copper. Plastic pipe is also usually marked with indelible writing. Plastic fittings will often have their identification cast into the fitting.

Water service pipes

Many materials are approved for water service piping. Some are more appropriate for underground or long installations. Materials will also differ greatly in cost. Before you lay water service pipe, remove rocks and sharp objects that can damage the pipe.

Preventing chemical reactions

Any pipe used for water distribution must be resistant to corrosive action caused by water. When installing potable water pipe below ground there are some special considerations. The pipe cannot be installed in soil that is contaminated with solvents, fuels, or other materials that can cause the permeation, corrosion, degradation, or structural failure of the pipe. If it is suspected that these detrimental conditions exist, a chemical analysis of the soil must be performed. If the location is deemed unsuitable for the installation of water piping, an approved alternative material or a different route will be required.

Potable water pipe and lead

The materials used for a potable water supply system may not contain more than 8 percent lead. This includes valves, faucets, and all other related materials used in the water distribution system. When you buy solder for the copper pipe, insist on 95/5 or lead-free solder. Even though 50/50 solder is still sold, it is not approved for use on potable water systems. The high-lead, 50/50 solder is restricted to use in heating systems and other non-potable piping applications.

Water service pipe ratings

All water service piping installed underground or outside the dwelling must have a minimum working pressure of 160 pounds per square inch (psi) at 73.4 degrees Fahrenheit. If the water pressure exceeds 160 psi, the pipe must be rated to handle the highest pressure possible from the water source. If you are using a plastic pipe for your water service, it must not extend more than 5 feet from the point of entry into the dwelling.

Approved water service pipes

There are numerous possibilities when selecting a water service pipe, but some are better than others. The materials discussed in this section are the most commonly used.

Brass pipe. Brass pipe is an approved material for water service use. If you are remodeling an old home, you might find a brass water service. While brass is still approved, it is rarely used in modern installations. The pipe is subject to corrosive actions and the threaded joints create the possibility of leaks below ground. I cannot think of any reason why you would want to use brass pipe in a new installation. There are many other approved materials that provide better service at a lower price.

Galvanized steel pipe. Galvanized steel pipe falls into a category similar to brass pipe. It is outdated and serves no purpose in modern plumbing installations. Galvanized steel pipe increases the risk of rust, leaking joints, and restricted water flow. Although steel pipe is approved, I cannot recommend its use as a water service pipe.

CPVC pipe. CPVC is approved for water service, but I don't like using it. Whenever feasible, it is best to eliminate joints in your underground piping, and with CPVC, you cannot eliminate the joints in long runs. Because most water service installations run for more than 20 feet, you will have underground connections with CPVC. Another concern with CPVC is its brittle construction. If the pipe is stressed during the backfilling of the ditch, it could break or the fittings leak.

If you decide to use CPVC, be especially conscientious when you lay and backfill the pipe. Remove any stones or objects that may break the pipe. When you are putting the pipe and fittings together be certain to work with clean, dry fittings and pipe. Always use cleaner and primer before you apply the glue. If your CPVC water line develops a leak after backfilling, you will have to dig it up again. This is not only aggravating, it gets expensive.

Polybutylene pipe. Polybutylene pipe is, in my opinion, a good choice for a water service pipe. The pipe is available in long rolls, eliminating the need for joints below the ground. Polybutylene pipe is flexible and resistive to damage from backfilling. If you get an unexpected, deeper-than-normal frost, polybutylene pipe will expand more than most pipes before it splits. Installing polybutylene is easy; you simply roll the pipe out in the ditch. When you compare costs, ease of installation, and effectiveness, polybutylene pipe is hard to beat.

Polyethylene pipe. Polyethylene pipe has long been used as a water service pipe; it is a proven performer. Polyethylene pipe also comes in large rolls, eliminating the need for underground joints. This plastic pipe is durable and resistive to backfilling accidents. With any water service pipe, you should remove all rocks and sharp objects that may puncture the piping. Polyethylene pipe is probably one of the most frequently used pipes for modern water service installations. It is inexpensive and easy to install.

Copper tubing. For many years, copper tubing was king of the water service pipes. It is still a fine choice, but it can be cost-prohibitive when compared to polybutylene or polyethylene pipes. Soft copper tubing, which is easily installed, is available in long rolls to eliminate underground joints. When you are choosing copper for your water service, you must decide what type of copper to use. Whether you choose type "L" or "K," you can be assured that copper tubing is a good water service pipe, unless the water supply contains a high acid content. The cost of the tubing will be considerably more than plastic pipe.

Water distribution pipes

In water distribution systems, the type of pipe used for hot water must also be used for cold water. Hot water pipe must have a minimum pressure rating of 80 psi at 180 degrees Fahrenheit. The rule pertaining to hot and cold water eliminates some of the pipes allowed as a water service pipe. The water distribution system will generally be comprised of one of the pipes in the following sections.

Brass pipe

Brass pipe is an approved water distribution pipe. Some older homes contain brass pipe throughout the water distribution system. As a modern plumber, you might connect to a brass system, but it is unlikely you will install one. Brass pipe must be threaded and screwed into fittings. This is time-consuming and unnecessary; there are plenty of viable materials that are less expensive and easier to install.

Aside from cost factors, brass is more likely to cause problems in later years. When the pipe is threaded, the walls of the pipe are weakened at the threads. These weak spots are an ideal place for corrosion to attack and cause leaks in the system. As with water service systems, I can think of no occasion where it makes sense to use brass pipe for a water distribution system.

Galvanized steel pipe

Galvanized steel pipe is an approved water distribution material, but I wouldn't use it. The pipe requires threading and screw joints. It is subject to restricted water flow from rust and leaks at the threads. Galvanized pipe, like brass, has been replaced by copper and plastic pipe.

Polybutylene plastic pipe

Polybutylene pipe is an approved material that makes sense. It is easy to install, inexpensive, resistive to splitting in freezing conditions, and flexible. This pipe comes in long rolls that can be pulled through a house in much the same way as electrical wires. The labor expended to install polybutylene pipe is minimal when compared to most other water distribution materials.

Polybutylene got some bad press when it was first introduced, but those problems seem to have been resolved. The special crimping tools needed to install polybutylene pipe are available at rental centers. For the average homeowner, polybutylene pipe offers an alternative to soldering copper pipe and fittings. Polybutylene pipe has been accepted by most professional plumbing companies as the pipe of the future. While old-school plumbers cling to copper, polybutylene pipe is paving its way into the houses of today and tomorrow. I think if more people knew about polybutylene, fewer would use CPVC pipe.

CPVC pipe

CPVC pipe is an approved material for water distribution systems and is favored by many homeowners. Homeowners like this pipe because they can glue it together and eliminate the need for soldering.

CPVC pipe has been around for many years, but it never has found favor with most professional plumbers. CPVC pipe has some advantages, but it also has disadvantages, especially for the professional plumber. One of the advantages is cost; CPVC pipe is less expensive than copper. For a professional plumber, the savings in material costs is lost in the time required to properly install it. CPVC pipe requires a cleaner, primer, and solvent to make an approved connection.

These three applications take time. In addition to the time lost in applying three solutions, the connection must set-up before it can be moved. All of this lost time makes CPVC pipe impractical for the professional plumber. When you acknowledge the fact that a pro's time is worth at least $30 an hour, watching CPVC cement set-up is an expensive pastime for professional plumbers.

If the joints between CPVC pipe and fittings are moved too soon, leaks are imminent. Also, if you drop CPVC on a concrete floor, it can crack. Often, these cracks go unnoticed until the water distribution system is tested. When the leak is discovered, you will have to cut out the leaking section and replace it. You must prevent water from running through the joints until the glue is dry. If water or dirt is present in the joints, they will leak. In cold weather, the drying time for the glue is extended, but if you have the time, CPVC pipe can get the job done.

CPVC pipe does not require any special installation tools. The pipe can be cut with a hacksaw, and the joints are made with a solvent. You do not have to master soldering skills to work with CPVC. The pipe is lightweight and inexpensive. Under average conditions, CPVC pipe can provide a long-lasting and easy-to-install water distribution system. If you don't mind slow working conditions, CPVC pipe is a viable choice.

Copper pipe

When copper pipe came onto the water distribution scene, it replaced galvanized steel pipe. Copper pipe did not have threaded joints to leak like galvanized pipe. Copper pipe does not develop rust and rough surfaces on the interior of the pipe. When compared to galvanized pipe, copper pipe offers many advantages. It is light in weight, easy to cut, easy to fit together, resistant to flow restrictions and leaks, and very durable.

Copper pipe has been the pipe of choice for water distribution systems over the last few decades. Copper pipe is probably found in more modern plumbing systems than any other type of water distribution pipe. While copper pipe is well regarded, it too has its deficiencies. If the potable water has a high acid content, the acid will react with the copper to deteriorate the pipe. Copper is expensive when compared to plastic alternatives.

For non-professionals, one reason not to use copper pipe is the need to solder joints. Soldering is not hard to learn, but it scares many people. Since soldering is the largest stumbling block for most people, let's look at how to solder a joint.

The first step in making a good solder joint is starting with clean fittings and pipe. You can buy wire brushes to clean the copper fittings. The pipe is cleaned with a fine-grit sandpaper. Brush the inside of the fittings until they are shiny and sand the ends of the pipe until they shine. The next step is the application of flux. Flux is a paste used during soldering to clean the pipe and fitting surfaces. Using a small brush, apply flux to the interior of the fitting's hub and to the end of the pipe. Do not allow flux to get into your eyes or mouth. Flux will cause a burning sensation and is not very tasty. Insert the pipe into the fitting until it is seated completely in the hub.

When you solder joints on potable water pipes, you must use a low-lead solder. In the old days, 50/50 solder was used, but not today. Buy 95/5 solder or lead-free solder for the joints. The old 50/50 solder is still available for making connections on non-potable pipes, but don't use it on potable water systems. With the joint fluxed and ready to solder, you will need solder and a torch.

After the torch is lit, hold the flame on the fitting's hub. You should wear safety glasses when soldering to protect your eyes from splashing solder and flux. As the fittings get hot, and they get very hot, you will see the flux start to bubble and drip. Lay the tip of the solder on the outside edge of the hub where it meets the pipe. It will take some time to become efficient with this part of the soldering process. If the solder doesn't melt and run around the pipe, remove it and continue heating the fitting.

Keep laying the solder on the joint to test the temperature. When the temperature is right, the solder will melt and run around the pipe. The heat will pull the solder into the fitting to make the joint between the fitting and pipe. Once you learn to gauge your heat temperatures, soldering is easy. Until you learn these temperatures, soldering can be frustrating. Let me give you a few pointers to reduce your frustration.

If the solder lumps up on the pipe when it is applied, the pipe and fitting are not hot enough to make a good joint. The lumping means the solder is being melted by the heat of the flame, not the heat of the pipe and fitting. A lumpy joint will normally leak. If the fitting turns black during heating, it is too hot. When the temperature is too hot, the joint will not make up properly and will probably leak. If the solder melts smoothly, but does not run around the pipe, you will have a leak. To cure this, apply more flux around the fitting and on the pipe. Be sure you are heating the pipe and fitting evenly and try the soldering again. If the solder still refuses to run, you will have to take the joint apart and clean the pipe and fitting better.

When the temperature is right, the solder will run smoothly around the pipe and fitting. Practice will allow you to become an expert at soldering. Small pipes, like 1/2-inch and 3/4-inch, will not take very long to reach their peak soldering temperature. Larger pipes, and brass fittings and valves will take longer to heat up. When you are soldering a joint at a valve, open the valve before soldering. The heat applied to the pipe will transfer to the washers in the valve and melt them if the valve is not open.

When you finish the solder joint, let it cool before you work with it. If you wiggle the pipe before the joint cools, it may leak. The cooling process doesn't take long. After the joint cools, wipe it with a cloth to remove excess flux. Flux left on the copper will turn green in a day or two. This makes for an ugly and unprofessional job. Refer to Figs. 2-1 through 2-5 for a step-by-step illustration of how to solder joints.

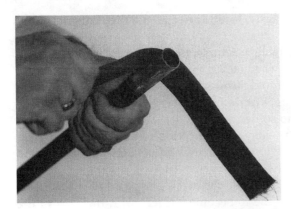

Fig. 2-1. Sanding copper pipe in preparation for soldering.

Fig. 2-2. Cleaning a copper fitting in preparation for soldering.

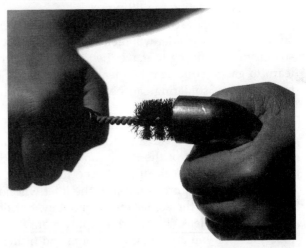

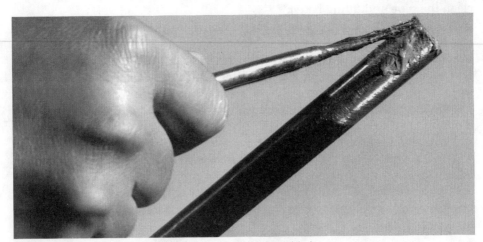

Fig. 2-3. Applying flux to copper pipe in preparation for soldering.

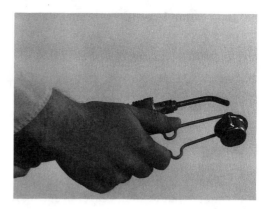

Fig. 2-4. Using a striker to light a torch.

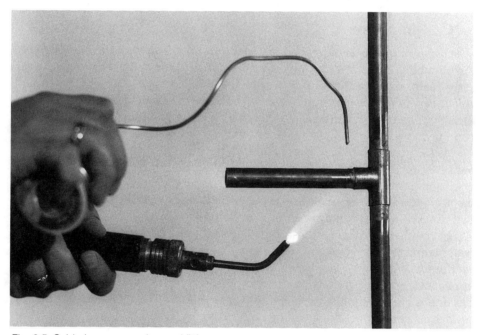

Fig. 2-5. Soldering copper pipe and fitting.

Copper pipe is durable and is approved for hot and cold water. Once you know how to solder, working with copper is easy. The pipe installs quickly and produces a neat looking job. Since the pipe is rigid, it may be installed so that the water system can be drained. This is a factor in seasonal cottages and other buildings where the water is turned off for the winter.

Copper pipe is used because it has become an industry standard. The public associates copper pipe with a quality job; they tend to look upon plastic pipe as second rate. While this perception may not be accurate, it is a fact. Public demand has much to do with the continued use of copper pipe.

What pipe would I use in my home? Being a middle-aged plumber, I would probably use copper, but I fully respect polybutylene. I can appreciate the many advantages polybutylene pipe has to offer, and I am open to changes for the better. As a homeowner, without soldering skills, I would use polybutylene pipe. The choice is yours, but you won't go wrong with copper or polybutylene pipe.

Above-ground drain and vent pipes

The plumbing code allows for the use of many types of pipes in above-ground drains and vents. While there is a large variety of pipes approved, only a few are commonly used. For example, aluminum tubing is an approved material for these applications, but I can never recall seeing it in use. Lead pipe is still a legal choice, but it is not used in modern plumbing.

Cast-iron pipe

Cast-iron pipe is an approved material for drains and vents. When I entered the plumbing trade, cast-iron pipe was the standard for drains and vents. In my early years, most cast-iron pipe was of the hubless type. The older, bell and spigot cast-iron pipe was rarely used. Today, both types of cast-iron pipe are found in homes with a little age on them. If you are planning a remodeling job, you may very well be tying your drains and vents into cast-iron pipe.

As far as new installations go, cast-iron pipe is rarely used in today's plumbing systems. Occasionally, cast-iron pipe will be used in multi-family properties or custom homes where drainage noise may be annoying. Some custom homes incorporate the use of cast-iron stacks to limit drainage noise, but in general, cast-iron drains and vents are a thing of the past.

In commercial jobs, cast-iron pipe is sometimes used to accommodate high water temperatures or unique conditions. For example, dishwashers in a restaurant produce extremely hot water. When plastic pipe is used as a drain for these appliances, the heat of the water may cause problems. Cast-iron and DWV copper drains are common substitutes for plastic in these situations. Under normal residential conditions, there will be no need to use cast-iron drains or vents.

Galvanized steel pipe

Galvanized steel pipe is an approved material for drains and vents, but it is not a logical choice. The pipe is expensive, it is labor intensive to install, and might be plagued with problems in the future. Galvanized drains are known for their problems with drain stoppages. As the pipe ages, it begins to rust on the inside. The rust creates rough surfaces that catch hair, grease, food particles, and other items, which block the drain. In time, these drains can close to the point that water will not pass through the pipe. With all the modern materials from which to choose, there is no reason to use galvanized pipe.

Brass pipe

Brass pipe is also approved for drains and vents, but it is an unlikely candidate for the job. When you compare the cost of the pipe, time to install the pipe, and the effectiveness of the pipe, it will be clear that you should make another choice.

Copper pipe

Copper pipe has been used as a drain-waste-vent (DWV) pipe for many years. It is a good pipe for the job, but it has become very expensive. As good as copper pipe is, it is not appreciably better than a schedule 40 plastic pipe for drains and vents. Unless you have money to burn, DWV copper pipe will be too expensive for serious consideration in your plumbing project.

Schedule 40 plastic pipe

Schedule 40 plastic pipe is by far the pipe of choice for modern DWV systems. Two types of this pipe are used for DWV systems: acrylonitrile butadiene styrene (ABS) and polyvinyl chloride (PVC). The most noticeable difference between these two types of schedule 40 plastic pipes are their colors. PVC pipe is white; ABS pipe is black. In addition to color, there are some other differences you will notice when you work with the two types of pipe. PVC is a more rigid pipe than ABS and is harder to cut. Presently, PVC is used more often than ABS. PVC is less expensive than ABS, but a little harder to install.

There are pros and cons to each pipe, but my personal preference is ABS. Why do I like ABS? When plastic pipe became the normal DWV pipe, I used ABS. I continued to use it for many years and developed great confidence in it. On some occasions, I used PVC at my customers' request. In these installations, I noted that there were some differences that convinced me that ABS was a better choice than PVC.

If PVC pipe is dropped on a concrete floor in cold weather, it may crack or shatter. I have had occasions when PVC pipe was dropped and seemed okay, until the completed system was tested. It is very discouraging and time consuming to find cracks in the piping after it is installed. The brittle nature of PVC pipe in cold weather is only one of the reasons I prefer ABS. ABS is almost indestructible; I have seen trucks run over ABS pipe without breaking it. When teaching apprenticeship classes, I have demonstrated the strength differences between PVC and ABS pipe by hitting each pipe with a heavy hammer. The PVC pipe breaks into pieces, while the ABS pipe is not phased by the blow. This durability factor is the first reason why I feel ABS is the better choice.

Working with both pipes over the years, I have experienced many more leaks with PVC than I have with ABS. PVC pipe is finicky; its joints must be dry, clean, primed, and glued under ideal conditions to avoid leaks. ABS almost never leaks. I have worked with ABS pipe in the rain and mud, without leaks. Installing PVC pipe under similar conditions almost would guarantee leaks. As I said earlier, ABS pipe is easier to cut than PVC. This is a small consideration, but if you spend much of your life cutting pipe, the easier the pipe is to cut, the easier your job. The easier cutting also translates into finishing the job faster.

ABS pipe is more flexible than PVC. This can make a big difference in confined working spaces. The ABS pipe can be bent to fit into a space where PVC will not go. A seemingly simple aspect of the two pipes can make the difference between getting the pipe installed, or not. There is a negative side to the flexibility of ABS pipe. If the pipe is left on a pipe rack, in the hot sun, the pipe will lose its form. The pipe will sag and develop a belly. This can cause problems in the drains, due to the low spots in the pipe. The negative effect of sagging may be avoided by keeping the pipe on a flat surface or out of the hot sun.

In the last few years, PVC pipe seems to have become the more common of the two pipes. In Maine, ABS is an oddity that is not stocked by most plumbing suppliers. Given a choice, I would choose ABS pipe, but PVC is also a fine DWV pipe.

Underground drain and vent pipe

When you install DWV pipe underground, the list of approved materials is not as long as the one for above-ground installations. Galvanized steel pipe, lead pipe, aluminum tubing, brass pipe, type "M" copper, and type "DWV" copper are all approved for above-ground use, but none of these pipes are allowed in underground installations.

Copper pipe

If you wish to use copper pipe for an underground drain, it must be either type "L" or "K" copper. There is no logical reason why anyone would use copper as a drain or vent in most modern plumbing systems. The only reasons I can think of would be if the pipe was carrying waste with chemicals in it or water at very high temperatures.

Cast-iron pipe

Cast-iron pipe is approved for underground use and is still used by some plumbers for underground plumbing. A few plumbers believe that cast-iron pipe will stand up to underground conditions better than schedule 40 plastic pipe. Personally, I can see little reason to use cast-iron pipe under average conditions.

Schedule 40 plastic pipe

ABS plastic pipe and PVC plastic pipe are both approved for underground plumbing applications. These schedule 40 plastic pipes are the most commonly used materials for drains and vents in modern plumbing applications. Except in special circumstances, either of these plastic pipes will do an admirable job in your plumbing project.

Sewer pipes

When it comes to sewer pipe, you may use any of the pipes described for underground piping. There are some other approved materials, but for standard applications, the pipes described above are the only ones with which you need to be concerned.

Fittings, valves, and nipples

All fittings, valves, and nipples used in a plumbing system must be made of approved materials. Further, they must be approved for use with the type of pipe being installed. Fittings shall not be made in a way that will restrict the flow of the pipe. All threaded drainage fittings must be of the recessed-drainage type.

Closet flanges

Water closet flanges must meet certain thickness requirements. Brass flanges must be at least $1/8$ inch thick. Plastic flanges must have a minimum thickness of $1/4$ inch. The flanges must be secured to the floor with corrosion-resistant screws or bolts. The bolts or screws used to mount the toilet to the flange must be made from brass.

Clean-outs

Clean-outs may be made from plastic or brass. Brass clean-outs may not be used with plastic pipe. Clean-out plugs must have a raised, square head, except where such a head would cause a safety hazard. In the event a raised head is unsafe, a countersunk head may be used.

Fixtures

Plumbing fixtures must have smooth, impervious surfaces. The fixtures must not contain defects or concealed fouling surfaces. All fixtures must conform to the standards of the local plumbing code. Some standard materials used in the construction of approved fixtures are: fiberglass, vitreous china, stainless steel, and enameled cast-iron.

Part II

Planning, installation, and inspections of new plumbing jobs

3

Blueprints and plumbing designs

IF YOU INSTALL your own plumbing you should be able to read blueprints. You must be prepared to design the plumbing system to conform with the local plumbing code. For most residential jobs these requirements are not beyond your reach. The average house is plumbed using only a small portion of the plumbing code, but the code must be followed accurately.

If you fail to install the plumbing in compliance with the local code, you may have to tear it out and replace it. This can be very frustrating and expensive. It is much easier to consider code complications in the design stage than it is after the pipes have been installed and a rejection slip issued.

Residential blueprints rarely show the plumbing design or where the pipes are to be installed. When you look at the blueprints for your home you may experience either of two feelings. The first may be that you have no idea where to begin. The second is that you may draw the plumbing on the plans anywhere you see fit. If you feel you have no idea where to begin, this chapter will help you. If you feel you can draw the plumbing in at will, you will be sadly surprised when you try to make the installation. What works on paper will not necessarily work in the field. This chapter is dedicated to making you aware of how to read blueprints and how to design a simple residential plumbing system.

Three major plumbing codes

There are three major plumbing codes used in the United States: the BOCA National Plumbing Code, the Uniform Plumbing Code, and the Standard Plumbing Code. There are other codes used, but these are the three major codes. The Standard Code is used primarily in the southern states. The Uniform Code is most prevalent in the western United States. The BOCA National Plumbing Code is found along the East Coast to where it meets the Uniform Code in the Midwest States.

Before you can begin designing your plumbing system, you must determine the code with which you will be working. Local jurisdictions may honor a different code or may modify one of the major codes. Calling the local code-enforcement office is the best way to ascertain information on the plumbing code in your area. Since I am a master plumber in states that use the BOCA National Code, most of the examples in this book will be based on the BOCA Code.

Since plumbing codes are amended regularly, there may be changes in the code by the time you read this book. Only use this information as a guide and a starting point. Before doing the job, consult the local plumbing inspector. Plumbing permits are required typically for new plumbing installations. Most local authorities will issue a permit to a homeowner to do his own plumbing. If you plan to do plumbing on a property you do not or will not reside in, you will have a problem. Normally, only master plumbers are allowed to pull permits for work on property other than their own home.

Obtaining a plumbing permit

The requirements for permit acquisition will vary from jurisdiction to jurisdiction. In some areas, all you will need is a completed form detailing the number and type of fixtures you will be installing. In other areas, you will need a riser diagram and a detail of all the plumbing to be installed. A riser diagram is a line drawing that shows how the plumbing will be installed (Fig. 3-1). The plumbing

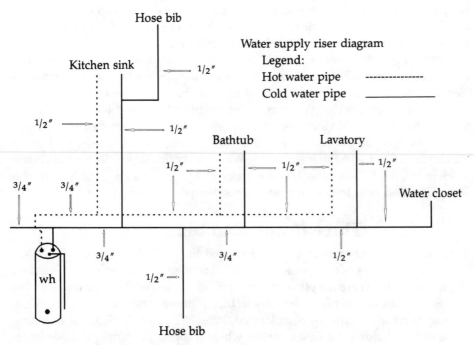

Fig. 3-1. A typical water supply riser diagram.

detail will show the proximity of the fixtures to walls and other fixtures. This is usually accomplished by providing a set of blueprints to the code-enforcement office. If your blueprints do not show the fixture locations, you will have to draw them.

An additional requirement may be a drawing that shows the location and method of installation for the water service pipe and the sewer pipe. If you are using a septic system, an approved septic design may also be required. You may need proof that you will be the resident of the dwelling if you don't hold a master plumber's license. Some code offices might require a list of the type of materials to be used in the installation. They might want everything from the type of pipe used to the type of fixtures you will be installing. All of these details should be decided when you design the system. The number from the building permit issued for the home might be required to obtain the plumbing permit. The last thing you will need is money to pay for the permit.

How to read blueprints

Residential blueprints are not very difficult to understand. They use common drafting symbols and are pretty straightforward. If there is a fault with residential blueprints, it is that they are too simple. In commercial plumbing, the blueprints are very specific. They detail everything you need to understand the plumbing layout. Residential prints rarely show more than the fixture locations for the plumbing. It is up to you to design the piping layout.

Symbols for residential plumbing fixtures

Even though you are only installing the plumbing, it is important to understand all aspects of the blueprints. For example, you might choose a route for a vent pipe that will be blocked by a recessed medicine cabinet. If you can read the plans, you can avoid these problems from the beginning. Let's take a look at some common symbols used in residential blueprints.

Bathroom groups. The first group of symbols we will look at are those found in bathroom groups. The three standard bathroom fixtures are: a water closet (toilet), a lavatory, and a bathing unit. To identify these items on your blueprints, refer to Figs. 3-2, 3-3, 3-4, and 3-5. The symbols on your blueprints may not match exactly these examples, but they should be close enough to enable you to identify easily the symbols on your blueprint.

If you are planning a double-bowl lavatory, it will look similar to Fig. 3-6. A corner shower will look like Fig. 3-7. A bidet is represented in Fig. 3-8. Soaking tubs and large whirlpool tubs are shown in Figs. 3-9 and 3-10.

Kitchen and laundry facilities. The kitchen will typically contain a kitchen sink and a dishwasher. Figure 3-11 shows a double-bowl kitchen sink and Fig. 3-12 indicates a dishwasher. Your laundry room plumbing will include a hook-up for a washing machine and might include a laundry tub. Figure 3-13 indicates a washing machine and Fig. 3-14 shows a single-bowl laundry tub.

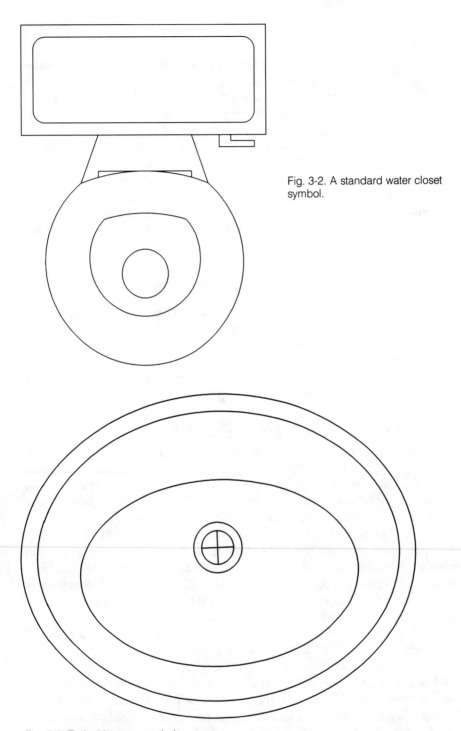

Fig. 3-2. A standard water closet symbol.

Fig. 3-3. Typical lavatory symbol.

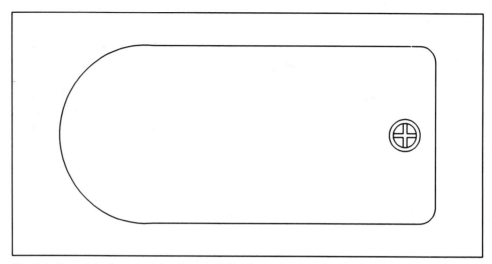

Fig. 3-4. Blueprints show the tub with a straight edge at the faucet end.

Fig. 3-5. The shower symbol uses an ''X'' to define the corners and center of the shower.

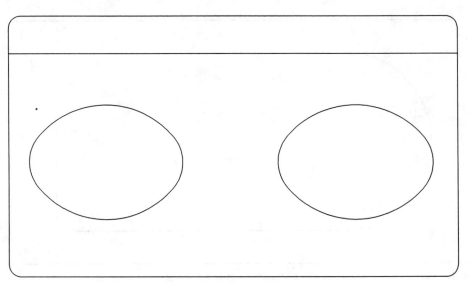

Fig. 3-6. Double-bowl lavatories are shown within the same rectangular top.

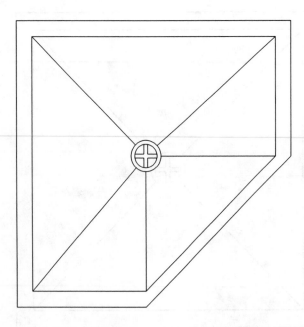

Fig. 3-7. The angled edge of a corner shower symbol shows the door/curtain location.

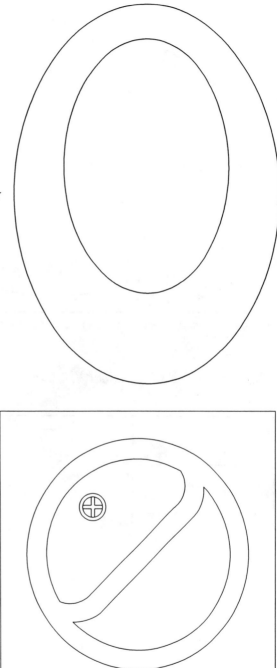

Fig. 3-8. The industry symbol for a bidet.

Fig. 3-9. A soaking tub symbol shows the unit's molded seat.

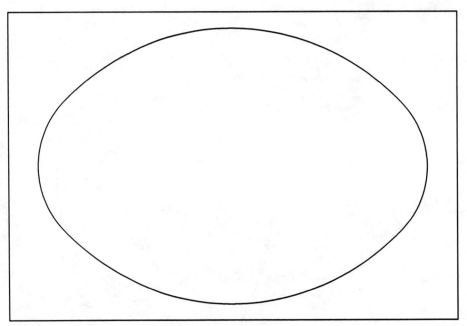

Fig. 3-10. The industry standard symbol for a whirlpool tub.

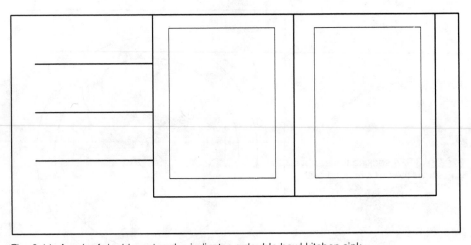

Fig. 3-11. A pair of double rectangles indicates a double-bowl kitchen sink.

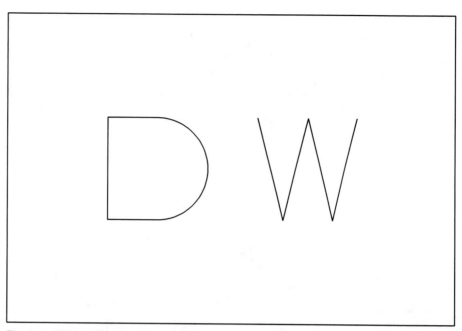

Fig. 3-12. DW indicates a dishwasher.

Fig. 3-13. An automatic washing machine is represented by either "AW" or simply "W."

Fig. 3-14. The standard laundry tub symbol.

Miscellaneous plumbing fixtures. Figure 3-15 shows a round water heater. Figure 3-16 represents a floor drain. A hose bib is shown in Fig. 3-17. A sump pit for an interior, waste-water pump is shown in Fig. 3-18.

Using these symbols as a guide, you should be able to spot the locations of all your plumbing fixtures on the blueprints. This is the first step in designing your piping layout. Once you know where all the fixtures belong, you must determine the best way to connect the fixtures to the plumbing system.

Fig. 3-15. Water heaters are usually shown as a labeled circle.

Fig. 3-16. The industry standard symbol for a floor drain.

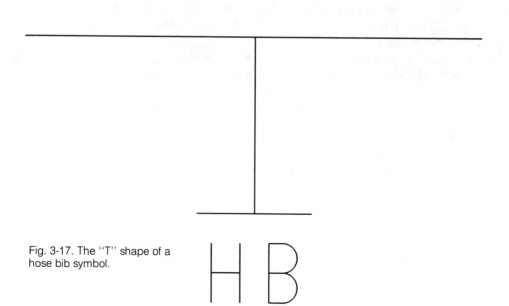

Fig. 3-17. The ''T'' shape of a hose bib symbol.

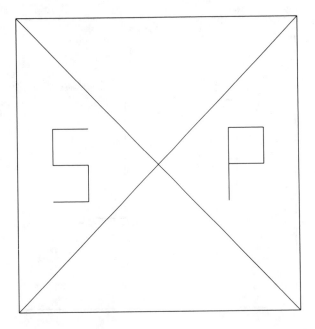

Fig. 3-18. The symbol for an interior waste-water sump pump.

Plumbing abbreviations

The following is a list of commonly used abbreviations for plumbing fixtures:

- CW—cold water
- SS—sanitary line
- FCO—floor clean-out
- HB—hose bib
- BS—bar sink
- WC—water closet
- EWH—electric water heater

- HW—hot water
- CO—clean-out
- WCO—wall clean-out
- FS—floor sink
- LAV—lavatory
- VTR—vent through roof
- GWH—gas water heater

Other blueprint items

It would take the space of another entire book to teach you every symbol used on blueprints. Since space is limited, I will give you an overview instead.

Several obstacles frequently affect the path of the plumbing in a house. Beams and floor joists always seem to be directly under the drains for toilets, showers, and bathtubs. When you read your plans, observe the floor joist detail. The joist detail will be on its own page in the framing layout. Measure on the actual plans to the centers of the drains to see if the drains will conflict with the joist or beam placement. When found in the planning stages, these obstacles are easily moved by the carpenters. If you don't discover the problem until after the house is framed, the relocation process becomes much more involved. The floor joists and beams are easily identified as the repetitive lines on the detail.

Steel plates and beams might cause you to change your plumbing plans. Observe the blueprints carefully to see if there is any steel in the path of your plumbing. When steel is present, it will usually be labeled on the prints. Notice the placement of windows and doors. Trying to run your pipes through these areas will not work well. Heating ducts are a major problem in some houses. Since it is easier to reroute plumbing than duct work, you will be the one that needs to make adjustments.

Electricians will normally work around your pipes. It is more reasonable for the electrician to pick a new direction than it is for the plumber. However, if you put a large pipe where the electrican must install a recessed switch or outlet, you will have to move it. The best way to avoid installation problems is to go over the plans and your proposed layout with the other trades involved in the installation.

Applying the code to your design

To gain a full understanding of the plumbing code is a monumental task. Code books are thick and cryptic, even for plumbers. I am going to give you pointers to get you started in the right direction, but don't depend solely on this information for your job. Codes differ and change. It would be next to impossible for me to give you specific details for installing your plumbing in compliance with the applicable code. Instead, I am going to show you some rule-of-thumb techniques that will pass most codes. Talk with your local plumbing inspector before using these examples.

In residential plumbing, you will be dealing with drains, vents, and water distribution. Learning code requirements for the water distribution system in residential plumbing is not extremely difficult. My methods will make sizing the pipe fairly simple. Pipe installation is governed by very few rules. The drain-waste-vent (DWV) system is a different story. The rules affecting the DWV system are much more complicated, but don't despair, I am going to simplify them for you. Installing the water service and sewer is relatively easy to grasp. Let's start with the water distribution system.

Code considerations for the water distribution system

The first factor to decide is the type of pipe and fittings you will use in the potable water system. The materials approved for this use were covered in Chapter 2. The next consideration is where you will run the pipes. This fact will be determined by the most economical layout and any obstacles in the path. Chapter 4 goes into detail on how to lay out the best route for your plumbing. For the design, pipe size is the most important consideration. You must determine what size pipe will be required to meet code requirements and to provide satisfactory service.

Sizing the water distribution system. In order to size your pipes, you need some benchmark information. This information will be found in the code book that applies to your area. These books should be available for purchase through the code-enforcement office. Since you may not have a code book or an understanding of how to use it, I will give you some examples.

Tables 3-1, 3-2, and 3-3 show the type of charts and information with which you will be working. Unless you are a plumber or a math wizard, the other sizing charts will only confuse you. Applying the charts shown in Tables 3-1, 3-2, and 3-3 and the others needed to complete pipe sizing is difficult even for experienced plumbers. In order to simplify the process, I am going to teach you a rule-of-thumb method for sizing residential water pipes.

**Table 3-1. Minimum sizes
for fixture water supply lines.**

Fixture	Pipe size (in inches)
Bathtub	1/2
Bidet	3/8
Dishwasher	1/2
Hose bib	1/2
Kitchen sink	1/2
Laundry tub	1/2
Lavatory	3/8
Shower	1/2
Water closet (two-piece)	3/8
Water closet (one-piece)	1/2

**Table 3-2. Fixture capacity
needs per square inch (psi)
and by gallons per minute (gpm).**

Fixture	Flow rate (gpm)	Flow pressure (psi)
Bathtub	4	8
Bidet	2	4
Dishwasher	2.75	8
Hose bib	5	8
Kitchen sink	2.5	8
Laundry tub	4	8
Lavatory	2	8
Shower	3	8
Water closet (two-piece)	3	8
Water closet (one-piece)	6	20

Table 3-3. Fixture unit ratings for hot and cold lines.

Fixture	Hot	Cold	Total
Bathtub	3	6	8
Bidet	1.5	1.5	2
Kitchen sink	1.5	1.5	2
Laundry tub	2	2	3
Lavatory	1.5	1.5	2
Shower	3	3	4
Water closet (two-piece)	0	5	5

Table 3-1 shows the minimum sizes of supply pipes for specific fixtures. This chart is easy to understand and is a good starting point for you. For most fixtures, a 1/2-inch pipe should be run to the fixture's cut-off valve. For toilets and sinks, a 3/8-inch supply tube will normally run from the cut-off valve to the point of connection on the fixture or faucet. With tubs and showers, a 1/2-inch pipe should be run directly to the faucet. When sizing the supply pipe for a water heater, 3/4-inch pipe should be used.

If you are using copper pipe, it is technically called tubing, not pipe. Half-inch pipe is actually 5/8-inch tubing and 3/4-inch pipe is actually 7/8-inch tubing. In this book, as in the trade, I will refer to the tubing as pipe and the sizes as 1/2-inch or 3/4-inch.

Most plumbing pipe is measured by its outside diameter. One exception is the tubing used for a standard dishwasher connection. This tubing is measured by its inside diameter and is called 3/8-inch tubing. This might confuse you when you buy 3/8-inch supply tubes for faucets and toilets. The supply tubes are measured on the outside and are much smaller than the 3/8-inch tubing used with dishwashers.

Sizing water pipe for the average home. In the average home, there are rarely more than 2 1/2 bathrooms. There will be a kitchen sink and a dishwasher. In addition, the plumbing will include two hose bibs and a washing machine connection. The last major fixture will be the water heater. This type of house, and houses with less plumbing, are easy to size.

Under most codes, the water service to the house might be a 3/4-inch pipe, but a 1-inch water service pipe is a better choice. The larger water service pipe allows for future expansion and provides a higher volume of water to the interior system. Once the water service is in the house, 3/4-inch pipe may serve as the artery for the water distribution system. The 3/4-inch pipe should run to the inlet of the water heater and serve as the primary pipe for the branch pipes feeding the fixtures.

Branch supplies to feed the fixtures may all be 1/2-inch pipe for the fixtures found in an average home. As you reach the end of the plumbing, the last two fixtures may be fed from a 1/2-inch pipe. There should never be more than two fixtures receiving their water from a single, 1/2-inch pipe.

Following these rules will keep you in compliance with most plumbing codes. If your home has additional fixtures, you may have to run a 1-inch water service pipe and extend, for a prescribed distance, the 1-inch pipe as the artery. Since technical sizing is so complicated, consult with a master plumber or the plumbing inspector if you have a higher fixture count. You will see examples of this type of sizing when you reach the section of this chapter on riser diagrams.

Sizing the DWV system

Sizing the DWV system is easier than sizing the water distribution system. The charts are easier to understand and the average person can size the DWV system. Tables 3-4 through 3-11 are used to determine the proper sizes for traps, drains, and vents. To decide on the appropriate trap size for your fixtures, refer to Table 3-5. Table 3-8 will show you which fittings may be used for various changes in direction.

Table 3-4. Fixture drainage unit values.

Fixture	Drainage fixture unit value
Bathtub	2
Bidet	1
Dishwasher	2
Kitchen sink	2
Laundry tub	2
Lavatory	1
Shower	2
Water closet (two-piece)	4
Water closet (one-piece)	4

Table 3-5. Minimum trap size requirements for various fixtures.

Fixture	Minimum trap sizes (in inches)
Bathtub	$1^1/_2$
Bidet	$1^1/_4$
Dishwasher	$1^1/_2$
Kitchen sink	$1^1/_2$
Laundry tub	$1^1/_2$
Lavatory	$1^1/_4$
Shower	2
Water closet (two-piece)	internal trap in fixture
Water closet (one-piece)	internal trap in fixture

Table 3-6.
Number of fixtures
allowed by drain pipe size.

Pipe size (in inches)	Number of fixtures allowed on pipe
2	21*
3	42**
4	216

NOTE: This table is based on the drainage pipe falling a quarter of an inch for each foot it travels. For example, a 4-foot length of pipe would be 1 inch lower at the discharge end than at the inlet point.

* Any pipe with a toilet connected to it must have a minimum size of 3 inches.

** Not more than two toilets or bathroom groups can be on a 3-inch pipe.

Table 3-7.
Pipe size requirements
for horizontal drain lines.

Pipe size (in inches)	Number of fixtures allowed on pipe
1 1/2	3
2	6
3	20*
4	160

* No more than two toilets or bathroom groups can be on a 3-inch pipe.

Table 3-8. Allowable fittings to accommodate changes in direction.

Type of fitting	Horizontal to vertical	Vertical to horizontal	Horizontal to horizontal
Sixteenth bend	yes	yes	yes
Eighth bend	yes	yes	yes
Sixth bend	yes	yes	yes
Quarter bend	yes	no	no
Short sweep	yes	yes	no
Long sweep	yes	yes	no
Sanitary tee	yes	no	no
Wye	yes	yes	yes
Combination wye and eighth bend	yes	yes	yes

To size the pipe, you must be able to calculate the number of fixture units the pipe will carry. Table 3-4 gives you the fixture unit ratings for common, household plumbing fixtures. To help you use these charts, let's count the fixture units for a one-bathroom home. The bathroom has a bathtub, lavatory, and toilet. The total fixture units for these fixtures is seven. The kitchen has a sink and a dishwasher. The fixture unit count for the kitchen is four. When we add the kitchen and bathroom figures, we arrive at a total fixture count of 11. If the home had two identical bathrooms, the total count would be 18.

Sizing the building drain and sewer. Look at Table 3-6 for the information you will need to size the drain. A 2-inch drain will not work because the home has a toilet in it. For a home with up to two bathrooms, a 3-inch building drain and sewer may be used. If you have more than two toilets in the home, a 4-inch pipe must be run for the sewer. The building drain must be a 4-inch pipe until the point where it only serves two toilets. This table is easy to use and shows that a 4-inch drain will be a very safe choice for the building drain and sewer. If you run a 3-inch sewer, you limit your ability to add additional plumbing at a later date.

Sizing horizontal fixture drains. When you know the fixture unit value, Table 3-7 will show you what size pipe is required for horizontal fixture drains. Checking Table 3-5 will also give you a good idea what size pipe is needed for various fixtures. The size of plumbing drains may be increased as they leave a fixture or trap, but they may not be reduced.

Sizing vents. Table 3-9 provides information for sizing vents to serve $1^1/_2$-inch and 2-inch drains. Table 3-10 provides information for vents used with 3-inch drains. Table 3-11 details vent sizing for 4-inch drains. These tables are also easy to use. Let me give you some examples.

Table 3-9. Size requirements for $1^1/_2$-inch and 2-inch vents.

Drain pipe size is $1^1/_2$ inches

# of fixtures on pipe	Vent size (in inches)	Maximum vent length (in feet)
1	$1^1/_4$	50
8	$1^1/_2$	150
10	$1^1/_4$	30
10	$1^1/_2$	100

Drain pipe size is 2 inches

# of fixtures on pipe	Vent size (in inches)	Maximum vent length (in feet)
12	$1^1/_4$	30
12	$1^1/_2$	75
12	2	200
20	$1^1/_4$	26
20	$1^1/_2$	50
20	2	150

Table 3-10. Vent size requirements for 3-inch drain lines.

Drain pipe size is 3 inches

# of fixtures on pipe	Vent size (in inches)	Maximum vent length (in feet)
10	1 1/2	42
10	2	150
10	3	1,040
21	1 1/2	32
21	2	110
21	3	810
53	1 1/2	27
53	2	94
53	3	680
102	1 1/2	25
102	2	86
102	3	620

Table 3-11. Four-inch drain vent requirements.

Drain pipe size is 4 inches.

# of fixtures on pipe	Vent size (in inches)	Maximum vent length (in feet)
43	2	35
43	3	250
43	4	980
140	2	27
140	3	200
140	4	750
320	2	23
320	3	170
320	4	640

If you have a 1 1/2-inch drain carrying a total of 10 fixture units, you may run a 1 1/2-inch vent for up to 150 feet. When figuring the main vent for your home, a 3-inch vent always should be adequate. Even if the charts indicate you could use a smaller vent, you must have at least one 3-inch vent penetrating the roof.

In our two-bath house, we had a total fixture count of 18. Table 3-10 shows that a 3-inch drain with 18 fixture units may have a 3-inch vent extending up to 810 feet. Even with 102 fixture units, you could extend the vent up to 620 feet.

Rule-of-thumb sizing. For a home, you can use some basics to design the DWV system. Use a 4-inch sewer and a building drain that is at least 3 inches in diameter. If you plan to have more than two toilets, the building drain will have to be a 4-inch pipe until there are only two toilets left to serve. For kitchen sinks, laundry tubs, lavatories, bidets, dishwashers, and bathtubs, run 1½-inch drains. In some codes, if you are using a dishwasher you may be required to run a 2-inch drain to the kitchen sink's trap arm. Bathtubs with a shower head above them are counted as bathtubs, not showers. For showers, run a 2-inch drain. Toilets should be served by no less than a 3-inch drain.

When you figure your vents, count on at least one 3-inch vent going through the roof. You will always be safe using the same size pipe for the vent that you use for the drain. In most cases, the vent may be a size smaller than the drain, but if you want to keep it simple, run the same size pipe for both the drain and the vent.

Riser diagrams

Riser diagrams are line drawings that show detailed information about a plumbing design. The riser diagram will show pipe sizes and the method of installation. These diagrams are usually required to obtain a plumbing permit. Very few residential blueprints supply a detailed plumbing diagram. You may have to draw your own riser plan for the code office. Drawing the detail will not be difficult if you refer to the example in Fig. 3-19.

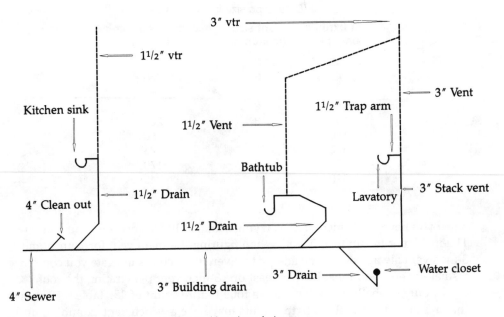

Fig. 3-19. DWV riser diagram, with size and location of pipes.

Once you have sized your piping, you will need to determine how you will install it. The next chapter will help you decide the best and most economical way to lay out your plumbing. Once you know the sizes and the layout, you are ready to draw your riser diagram. Look over the example in Fig. 3-19. There are additional examples in the next chapter. Read Chapter 4 and then try your hand at drawing a riser detail for your project. It will not be as difficult as you might think.

4

Layouts and material take-offs

THIS CHAPTER DEALS with job-site layouts and material take-offs. While designs on paper are needed, you ultimately work with the job-site layout. If you wish to save money on the plumbing layout, you must look ahead before you lay out the job. When you save 20 feet of copper, you save money. This chapter will show you what to look for and how to minimize the materials needed for your plumbing job.

Plumbing designs are done in the planning stage of a job. They show an intended layout for the plumbing. In large commercial jobs, the plumbing is usually installed as it was drawn in the design. Residential plumbing almost never goes according to the original plan. Since residential blueprints are not as thorough as commercial prints, the job has more latitude for change.

The blueprints for an average house will be detailed for the carpenters, but leave much to be desired for the plumber. Occasionally, there will be a section dealing with the heating and electrical systems, but it is rare that you will find a plumbing page among the blueprints. When you make your plumbing design on paper, you have no way of knowing where the other trades will be placing their work. Heating ducts, wires, floor joists, and a host of other unexpected obstacles might force you to change your planned design.

Remodeling jobs are even more apt to require on-site adjustments to the intended design. Whenever you are working with concealed piping and existing conditions, changes to the design are likely. Even if you are working with an engineered design, it may not do you much good on residential work. The lack of conformance to plans in residential work is common and accepted. What does all this mean to you? It means you must be prepared to make on-site changes in your design. You may not go into residential plumbing with the idea that the job will be installed exactly as it was drawn in the design.

Sizing pipes

Learning to size the piping is the first step to laying out the plumbing design. It is also a key element in controlling the cost of the job. Figure 4-1 illustrates an inefficient water supply layout. Running a larger pipe than needed means spending more money than necessary. Installing oversized pipe is wise if you anticipate expanding the plumbing system later, but there are areas of the system where oversized pipe is a waste of money. A good example of this is the water pipes running to a shower valve.

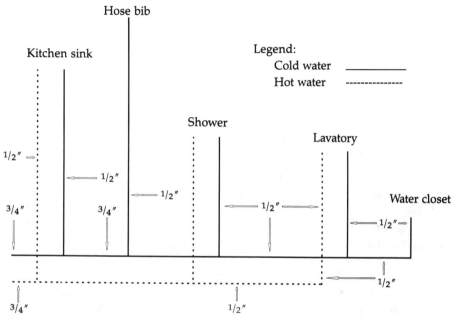

Fig. 4-1. Inefficient water supply design.

Shower heads are designed to produce a flow of 3 gallons of water per minute. A 1/2-inch pipe will provide an adequate supply of water to achieve the intended flow rate. If you run 3/4-pipe to the shower valve, you are wasting money. Let's say the shower is located 15 feet from the main water distribution pipes. This means you will run 15 feet of pipe for the cold water and another 15 feet for the hot water. At the time of this writing, the difference in cost between 15 feet of 1/2-inch copper and 15 feet of 3/4-inch copper is $9.30. This may not seem like a lot of money, but if you look at the cost of the complete job, these savings add up.

Consider the cost of running an oversized water service with copper pipe. Assume a 3/4-inch pipe is large enough to serve your home with incoming water from the street. In this example, the length of the water service is 75 feet. If you run 1-inch copper instead of 3/4-inch copper, the additional cost will be $27. Figure 4-2 is an example of an efficient water pipe design. The total length of pipe

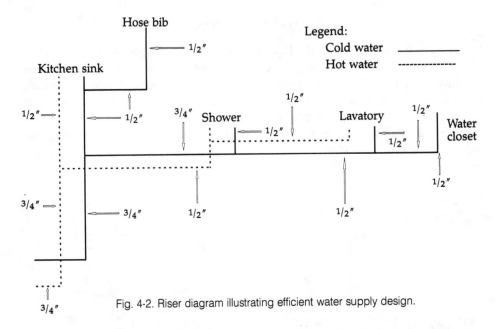

Fig. 4-2. Riser diagram illustrating efficient water supply design.

used is significantly reduced from the diagram in Fig. 4-1. Now let's look at an example for the drain-waste-vent (DWV) system.

After you have one 3-inch vent penetrating the roof, most other bathrooms may be vented with a 2-inch vent. You may also tie many individual vents into the main 3-inch vent. Assume you have a 2-story home with two bathrooms on the first floor. If you vent separately both of these bathrooms with a 3-inch vent, you are spending about $20.00 more than necessary. You are paying for a second 3-inch vent when it could be a 2-inch vent. An additional roof flashing is required and you are using much more pipe than if you tied the two vents together. Figure 4-3 illustrates an inefficient DWV riser plan. Note the number of vents penetrating the roof. The DWV riser diagram in Fig. 4-4 shows a more efficient and cost effective use of pipe.

These examples give you a good idea of how the pipe size and layout can affect the job's cost. Over the scope of the entire job, you might save hundreds of dollars with the right layout.

Cost of cutting out new plumbing

The two main causes for removing newly installed plumbing are code violations and conflicts with other trades. When you are forced to remove and reinstall new plumbing, you are losing money. There is an obvious loss of your time; you know how much your time is worth. In addition, there might be significant losses in wasted pipe and fittings. Once schedule 40 plastic pipe and fittings are installed in a DWV system, they are difficult to reuse. The pipe may not be removed from the fittings once the glue has set up. You might be able to piece the

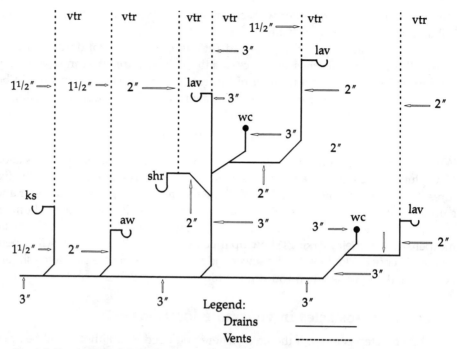

Fig. 4-3. Poorly designed DWV layout.

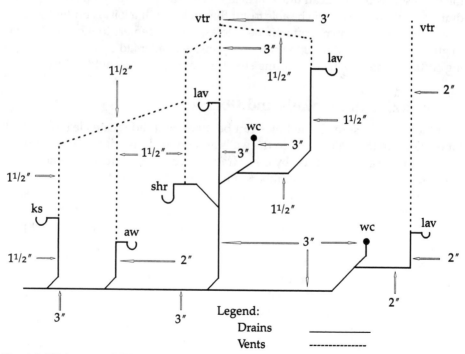

Fig. 4-4. Efficient use of DWV pipes.

fittings and pipe back together with couplings, but this is time consuming and still amounts to a more costly job.

If you have to replace the pipe and fittings used for a standard bathroom group, the cost in materials will exceed $15. This example assumes you only need to replace a minimum number of fittings. In some cases, the cost might be much higher. This type of expense may be avoided with the proper planning.

On-site layouts

When you begin your on-site layout, your first consideration should be the location of all plumbing fixtures. The blueprints should show the location for each of the fixtures. Go through the house and mark all the fixture locations on the sub-floor and stud walls. A lumber crayon is very useful for marking the fixture locations. Lumber crayons are available at most stores selling building or plumbing materials. They are inexpensive and their markings show up much better than those of a pencil or pen. Find and verify the location for the sewer and water service. These two items should be your starting points when you run the plumbing.

Look for obstacles in your hole locations

The next step is to see if the areas where you need to drill holes will be open for your pipes. Floor joists are often in the way of drains for tubs, showers, and toilets. Just because you can draw a circle on the sub-floor for your water closet's drain doesn't mean you will be able to cut the circle out. If a joist is under the floor, where the circle is drawn, you may have to ask the carpenters to move it. Or, you might cut it in half and have the carpenters install headers to support the remaining sections. Before you start cutting structural members, consult the carpenters.

Obstacles in the walls and floors

Once you are sure all the holes can be made with adequate clearance, look for obstacles in the walls. Look in the walls and on the blueprints. Avoid running pipes in areas selected for use by other trades. Recessed medicine cabinets are a frequent problem for novice plumbers. Plumbers often bring their lavatory drain up in the wall so that it is in the center of the lavatory. From the top of the sanitary tee, they extend the vent straight up. This is where they run into a problem. If the lavatory is to be equipped with a recessed medicine cabinet, the vent pipe will be in the way.

It is a safer bet to bring the drain up in the wall off to one side of the lavatory and run a horizontal pipe over to the center of the lavatory. This allows the vent to go straight up from the tee without a conflict. The vent will not be near the medicine cabinet and you will not have to move the vent later.

Look out for duct work locations if the ducts have not been installed already. A good builder will have the duct work installed before the plumbing, but there are times when the plumber goes to work first. See if the blueprints have a heating diagram. If there is no diagram, go over the heating layout with the appropriate person before you install the pipes. If you don't, the odds are high you will be relocating your work later. Apply the same principle to the electrical work.

Electrical work is less of a problem than heating systems, but electrical locations may prohibit you from using the space. Recessed lights, switches, outlets, and other related items might cause you to change your intended path. If the wiring has been installed before the plumbing, be careful with your drill. In most cases, if you drill through the wires you will be responsible for paying the costs to repair them.

Watch out for windows

When you select the locations for the vents that penetrate the roof, watch out for windows. If the pipe comes through the roof within 10 feet of any operable window, door, or ventilating device, it must extend until it is 2 feet above the window, door, or device. This makes an ugly installation and should be avoided whenever possible. When you are comfortable the holes will be clear and there are no major obstacles to avoid, you are ready to lay out the plumbing.

Laying out the piping

When you begin the piping layout, start with the DWV system. Water pipe requires less space and is easier to maneuver than the DWV pipes. Start your layout at the sewer. Make a mental diagram of how you will route the drains and vents. Be aware of beams, joists, pieces of steel, and other factors that might influence your decision. Look for ways to use the least amount of pipe. Keep in mind the amount of fall you will need for the proper grade on the pipes.

Many people forget that their drains will be falling at a rate of a $1/4$ inch for each 12 inches they travel. If you are working where all pipes must be contained in a ceiling, this might be a problem. If you have a drain running 20 feet along the basement wall, the low end of the pipe will be 5 inches lower than the other end.

When you have a handle on the DWV pipes, start on the water distribution pipes. Avoid running the pipes in outside walls and attics in cold climates. Consider the effect of wind drafts around basement windows and crawl-space vents. In cold climates, these factors contribute to frozen pipes. When you plan the water distribution system, look for ways to reduce the amount of pipe used. Refer to the riser diagrams in Figs. 4-1, 4-2, 4-3, and 4-4 as examples for routing the pipes for maximum efficiency and economy. When you have completed your on-site layout, make a simple drawing of how your plumbing will be installed.

Tips for economical layouts

It is difficult to give you precise examples of how to design the most economical plumbing system for the job. Each job has its own set of circumstances, which affect the layout. Since I can't give you specifics for your job, I will give you pointers for typical money-saving methods.

If the choice is yours, choose hose bib locations close to other plumbing fixtures. If you install the hose bibs near the kitchen sink, bathroom, water service, or water heater, you will save money. With these locations you may install a tee in one of the other water pipes to serve the hose bib. This eliminates the need for running extensive piping to the hose bib.

Position the water heater in a location near the bathroom. Since you may run $1/2$-inch pipe to two fixtures, you will save on the cost of $3/4$-inch pipe. In a one-bath house, the only two fixtures using hot water other than the bathroom will be the kitchen sink and a washing machine. You may run $1/2$-inch pipe to serve both of these fixtures.

Run $3/4$-inch pipe to the tub for the cold water and branch off this pipe with $1/2$-inch pipe for the lavatory and toilet. If you have a dishwasher, it may receive its water supply from the pipe serving the kitchen sink. This is a close run of pipe and reduces the amount of pipe used. When you have a refrigerator with an ice-maker, feed the ice-maker from the pipe serving the kitchen sink. This is usually done with a saddle valve and $1/4$-inch, plastic tubing.

When you install the vents for the DWV system, tie as many of them together as possible. Once the vent is 6 inches above a fixture's flood-level rim it may be offset. By using offsets, and connecting individual vents to a main vent, you will save pipe and money. Every time you eliminate a roof penetration, you save the cost of a roof flashing.

Avoid unnecessary turns in the drainage piping. When you eliminate sharp turns, you reduce the risk of drain stoppages. Gradual turns keeps the system running smoother.

Buy your pipe in 20-foot lengths. This will save you money in two ways: the pipe will be cheaper and you will use fewer couplings. Shopping for the materials may turn up some surprises. Buying the materials from the best supplier might save you 35 percent, or more.

Material take-offs

Once you have a detailed layout of the plumbing, making an accurate material list is easy. Look over your on-site diagram and count the number and type of fittings you need. Being on-site when you make the material list is helpful. You can make measurements for the amount of pipe you will need and see how the plumbing will actually be installed.

What should be included on a material take-off? Everything you will need to install the plumbing should be on the list of materials. Pipe, fittings, roof flashings, pipe hangers, and any other items should be on the list. I have provided forms to help you make a good material list. These forms are labeled as Tables 4-1, 4-2, 4-3, and 4-4.

Accurate material take-offs save you money and frustration. If you are installing plumbing and don't have a fitting you need, frustration will run high. If you are doing the job when stores are closed, your job will come to a rapid halt.

Table 4-1. Material take-off for DWV fittings.

Fitting	Size	Street/hub	Quantity
Coupling			
Wye			
Eighth bend			
Long-turn quarter bend			
Quarter bend			
Cap			
Sanitary tee			
Sixteenth bend			
Male adapter			
Female adapter			
Clean-out			
Clean-out plug			
Test tee			
Wye and eighth bend			
Closet flange			
Trap adapter			
P-trap			
Shower drain			
Reducer			
Reducing wye			
Reducing tee			
Tub waste/overflow			
Other			

When you have to leave the job site to pick up additional materials, you lose time and money in travel expense. A good list of materials will make the job go smoother and more economically.

Table 4-2. Material take-off for water pipe fittings.

Fitting	Size	Street/hub	Quantity
Coupling			
Cap			
Tee			
Reducing tee			
Male adapter			
Female adapter			
Union			
45 degree ell			
90 degree ell			
Drop-ear 90 degree ell			
Stop/waste valve			
Gate valve			
Hose bib			
Bushing			
Other			
Other			
Other			

Table 4-3. Material take-off for pipe.

Size (in inches)	Type	Quantity
$1/4$		
$3/8$ id		
$3/8$ od		
$1/2$		
$3/4$		
1		
$1 1/4$		

Table 4-3. Continued.

Size (in inches)	Type	Quantity
1 1/2		
2		
3		
4		

Table 4-4. Material take-off for supplies.

Item	Size	Type	Quantity
Solder			
Flux			
Flux brush			
Cleaning brushes			
Sandpaper			
Cleaner			
Primer			
Solvent/glue			
Rubber couplings			
Clamps			
Pipe hangers			
Putty			
Pipe compound			
Saw blades			
Nails			
Roof flashings			

In conclusion

Your plumbing layout will influence and be influenced by many job factors. It will have an effect on the cost of the job and the performance of the plumbing. Spend some extra time and lay out an effective system before you begin. If you simply start hanging pipe, without a plan, you will find yourself in trouble before you know it.

5

Cost estimating and material acquisition

LEARNING TO estimate your material costs will allow you to budget for your plumbing projects. Once you have compiled an accurate material take-off, estimating your costs will be simple. Chapter 4 taught you how to make material take-offs. This chapter will teach you to estimate the cost of your materials. If you are doing a large plumbing job, you have to consider when to order various materials. If you order too soon, your materials may be stolen or damaged. If you wait too long to place your order, the job may be held up. Timing is an important element in material acquisitions.

The type of job will dictate when materials should be delivered. Remodeling jobs are less likely to be affected by theft and vandalism than new construction jobs. When people live at the job site, the risk of illegal activity is reduced.

New construction jobs are often set off by themselves. Without neighbors to keep an eye on the job site, conditions are right for criminals. If you are new to plumbing, there are some factors you might not think of without the following advice.

Job-site theft

Theft has probably always been a problem in the plumbing trade, but in the last few years it has become a big factor in the cost of plumbing. I started in the trade some 18 years ago. At that time, theft was hardly a consideration. In the last 10 years, I have seen job-site losses escalate. I'm not talking about losing a few copper fittings or a length of pipe. The theft you need to be most concerned about is the big money items.

Bathtubs and showers

New construction requires the installation of bathtubs and showers at an early stage. The house may not even be in a lockable condition when these bathing units are installed. It is also common practice to have the tubs and showers delivered to the job site and dropped off outside the house. Many homes include two bathing units. Tubs and showers cost a professional plumber around $300 each. If these units are stolen, the plumber is out $600.

The units are not safe even after they are installed. Most houses are equipped with one-piece, fiberglass tubs and showers. These units are installed by nailing or screwing them to the stud wall. It is a simple matter to remove the nails or screws. These are large items, but that will not deter a thief. Anyone with a pick-up truck can haul your tubs or showers away in a matter of minutes.

If the house can be seen by neighbors or passersby, night-time activity will attract attention. When these same citizens see a pick-up truck on the job during the day, they don't give it a second thought. They think the truck and its occupants are part of the crew working on the house.

Rainy days are a prime time for job-site crooks. Since many of the trades don't work in the rain, the site is left unattended. The thieves will drive up and load everything they want into their truck. I have had this type of theft occur even with my own men on the job. If the bad guys are questioned, they will say their boss sent them out to pick up whatever it is that they are stealing. If they are questioned beyond this point, they may plead innocence by reason of bad directions. They will attempt to convince the interrogator that they must be at the wrong house. Professional job-site pirates will have a long list of plausible excuses to get them off the hook.

Serious bad guys will follow delivery trucks to the job site. After the delivery is made, the crooks will sweep in and pick up the new materials. When the homeowner or contractor checks the job, he assumes the material was never delivered. This results in a conflict between the purchaser and the supplier. In these cases, the theft may not be identified as a theft for several days. By the time it is determined the material was delivered and stolen, the goods are long gone.

Copper pipe and fittings

If you leave pipe and fittings at the job site, they may not be there when you go back. If you are only going to lunch, the material may disappear while you are gone. Even if you guard against this type of theft, the pipe may be stolen after it is installed. Copper is bought in scrap yards and brings good money by the pound.

In urban areas, it is not unusual to have all the copper plumbing cut out by thieves and sold for scrap. On the surface, this may seem like a lot of work and risk for the reward, but it happens. I have installed copper on a Friday and had it stolen before I returned on Monday. This type of theft is not as common in rural areas, but it is a risk to be considered.

Fixtures

Trim-out fixtures are ordered when the job is coming to an end. When these fixtures are delivered, they are often set in the garage or just inside the house. The cost of these fixtures may easily exceed $1,000. If the house is not secure, these items are attractive to the criminal element. In less than 15 minutes, a professional can load and leave with these high-dollar items.

Avoiding job-site theft

There will always be a degree of risk related to theft, but you can reduce the risk with a few standard procedures. If you are plumbing a new house, don't order the tubs and showers until you are ready to install them. If possible, wait until the house can be locked to install the bathing units. Arrange for deliveries when you are there to accept them. If you are there to take delivery, you are sure the materials were delivered and in good condition. Don't leave materials at the job site unattended.

If the job site is at the end of a long driveway, put a cable or chain across the driveway when you are not working. Give each of the other authorized workers a key, but insist they keep the barrier in place when they are not on the job. Criminals will not want to carry heavy fixtures up a long driveway to steal them. If you keep their truck at a distance, you reduce the odds of a major theft.

When you have a friendly neighbor, talk to them about your job site. Describe the type of vehicles that will be going in and out of the site. Ask the neighbor to notify you or the police if strange vehicles are present on the job. Talk to the other tradespeople and ask them to keep an eye out for unknown people and vehicles. Sometimes a simple sign will protect your job. Place a sign at the driveway and on the home indicating the job is patrolled by a private security agency. Even if the sign is fictitious, the thieves may believe it and move on to another site.

Once the job site can be locked, keep it locked. If it is your house, you may have to go to the job each evening to lock it up. Not all contractors are reliable when it comes to job-site security. Use common sense and take these practical precautions to protect your job. Criminals usually seek easy targets. If your job intimidates them, they will move to a job with less security precautions.

Cost estimating

If plumbing is only an occasional part of your life, cost estimating will be a little more difficult than it is for a professional. Professional plumbers have complete files and databases of information at their disposal for estimating a job's cost. As a part-time plumber, you will have to work a little harder to develop an accurate cost estimate.

After you have a plumbing design and a material take-off, estimating is only a matter of assigning a value to the materials you need. The easiest way for a homeowner or part-timer to do this is to work with a material supplier. Take your take-off list to the supplier and ask for a price quote. This is not only the easiest way, it is also the most accurate method of cost estimating.

When a supplier gives you a price quote, it will usually be good for between 10 and 30 days. Preliminary estimates are often made long before the job is started. Since the quoted price will expire before the work is started, the prices are only estimated costs for the job. When you work up your estimate, allow for changes in the design. You learned in Chapter 4 how your plumbing plans can change quickly. These changes will affect your material list and estimated cost.

As you plan your budget, add a minimum of 10 percent to the estimated job cost. This 10 percent will offset price increases, changes in plans, and forgotten materials. If you are new at plumbing, add 20 percent to the estimated cost. As a professional, I add 7 percent to my anticipated costs. Without experience in doing take-offs, 20 percent is a safer figure for you to use.

If sales tax is applicable in your location, don't forget to allow for it in your estimate. At discounted prices for professional plumbers, the materials for a $2^1/_2$-bath house may cost more than $2,500. The sales tax on this amount of money can make a substantial difference in your job's cost.

Will you have electrical power available to do your work? If power is not available on the job, you will need a generator. Generators may be rented, but they add to the cost of the job. Plumbing permits are another soft cost for which you must plan. Permits are not major expenses, but they might cost $50 or more. Will you be losing paid time from your regular job to do the plumbing? If you are giving up income from your full-time job to do the work yourself, factor in the lost income as a part of your plumbing expense. Even insurance premiums may add to your plumbing costs. Any cost you might incur when installing the plumbing should be considered a part of the job's cost.

Price quotes

When the time comes to purchase your plumbing materials, where and how you purchase them may influence the cost of the job. After being in the plumbing business for many years, I have learned the hard way about buying materials. This is an area where many people, including professionals, waste large sums of money. I am not suggesting you use inferior materials to save money, but with the proper methods, you can see significant savings through aggressive shopping.

As the time to perform the work draws near, put your material list out to bid. Take identical copies of your material list to numerous plumbing suppliers. Be sure each item on the list is clearly specified. When you specify the materials, include model numbers, sizes, brand names, and the type of material. If you are not extremely specific, you will not receive fair quotes.

Experienced suppliers are adept at making their price look the best. Never look only at the bottom line. It is unusual when one supplier can offer you the best price on all the different types of material. One supplier may have the best price on fixtures, but be high on the prices of pipe and fittings. To effectively shop, you must make sure your comparisons are exact.

If the supplier throws in a clause that states he will supply the specified material, or an equal material, you have the makings for trouble. Who decides what a fair equal is? In these circumstances, the supplier makes the decision. You are paying the bill and should be allowed to make decisions on any changes.

The "or equal" clause is only one way suppliers will camouflage their pricing. Some suppliers will intentionally delete items from their quote to have the final number be lower than the competition. If you do not compare the bids item-for-item, this sales tactic may go unnoticed. When you are soliciting bids you must be thorough and cautious. Look for any discrepancies between the bids and take note of how long the prices are guaranteed. Some quotes will only be good through the day they are given. Others will remain in force for 30 days.

Cutting the best deal on materials

Local suppliers are not always the best choice. If you live in a small town, send quote requests to the nearest city. When competition is stiff, prices are generally lower. Cities support more suppliers and the prices offered are frequently to your advantage. I live in Maine, but I can save 35 percent on many of my materials by purchasing them in Massachusetts. This type of savings gives me an edge in a competitive market place. I can bid the job for much less than the plumbers who buy materials locally and still make more money. As a homeowner, you, too, can use these savings.

As you narrow down the list of potential suppliers, get all your figures lined up on a spreadsheet. Approach each of the suppliers and tell them they are in the ballpark, but need to shave a little off their prices. If the supplier wants your business, he will reach into his buffer zone and offer a lower price. These tactics will take a little of your time, but they might save you substantial money.

The benefits

The benefits offered from the information in this chapter are many. By following this advice you should save money when you purchase materials, and avoid on-site losses. The extent to which you wish to save money is up to you, but these suggestions give you the knowledge on which to build a strong foundation of savings. With the proper application, these suggestions may make all of your plumbing jobs more enjoyable and more profitable.

6

Installing underground plumbing

UNDERGROUND PLUMBING is frequently overlooked in books that deal with plumbing. While underground plumbing receives little recognition, it is instrumental in the successful operation of your home's plumbing. Your home might not have any interior plumbing below grade, but it will have a sewer underground. If the sewer fails to perform, the remainder of your drainage system will not operate.

Underground plumbing is installed before the house is built. If you will have interior underground plumbing, it will be installed before the concrete floor is poured. When the plumbing is not installed accurately, you will have to break up the new concrete floor to relocate it. Since the placement of the underground plumbing is critical, you must be acutely aware of how it is installed. Miscalculating a measurement in the above-grade plumbing may mean cutting out the plumbing, but the same mistake with underground plumbing will be more troublesome to correct.

Underground plumbing is often called *groundwork* by professional plumbers. The groundwork is routinely installed after the footings for the home are poured and before the concrete slab is installed. When you install underground plumbing, you will have to take note of code differences between groundworks and above-grade plumbing. For example, while it is correct to run a 1½-inch drain for an above-grade bathtub, you must run a 2-inch drain if it is under concrete. Let's take a moment to look at some of the code variations between interior and underground plumbing.

Code considerations

Underground plumbing is quite different than above-grade plumbing. Several rules pertain to groundworks that do not apply to above-grade plumbing. This is true of both the drain-waste-vent (DWV) system and the water distribution system. The following information will highlight the common differences. This information may not apply in your jurisdiction and the information is not inclusive of all code requirements. The following code considerations are the ones most often encountered with underground plumbing.

DWV systems

The differences for DWV installations are minimal, but they must be observed. Galvanized steel pipe is an approved material for the DWV system when the pipe is located above grade. But galvanized drain pipes may not be installed below ground. This rule has little effect on you since most DWV systems today are plumbed with schedule 40 plastic pipe.

The minimum pipe size allowed under concrete is 2 inches. This rule could cause the unsuspecting plumber trouble. When you check your sizing charts and see that $1^1/2$-inch pipe is an approved size, you might choose to use it under concrete. If you do, you will be in violation of most codes. While $1^1/2$-inch pipe is fine above grade, it is illegal below grade.

The type of fittings allowed are also different. When working above grade, a short-turn quarter bend is allowed when you change directions from horizontal to vertical. This is not true below concrete. Long-sweep quarter bends must be used for all 90-degree turns under the concrete.

When you run drains and vents above grade, the pipes must be supported every 4 feet. This is done with some style of pipe hanger. When you install pipes underground, you will not use pipe hangers. Instead, you will support the pipe on firm earth or with sand or gravel. The pipe must be supported firmly and evenly.

When the groundworks leave the foundation of your house, the pipe will have to be protected by a sleeve. The sleeve may be a piece of schedule 40 plastic pipe, but it must be two pipe sizes larger than the pipe passing through the sleeve. This rule applies to any pipe passing through the foundation or under the footing.

It will be rare that any residential drains under concrete will run for 50 feet, but if they do, you must install a clean-out in the pipe. Pipes with a diameter of 4 inches or less must be equipped with a clean-out every 50 feet in a horizontal run. The clean-outs must extend to the finished floor grade.

As your building drain becomes a sewer, you must adhere to yet more rules. The home's sewer must have a clean-out for every 100 feet it runs. There should also be a clean-out within 5 feet of the house where the pipe exits. These clean-outs must extend to the finished grade. If the sewer takes a turn of more than 45 degrees, a clean-out must be installed. Clean-outs should be the same size as the pipes they serve. For the drainage system, clean-outs must be installed so that they will open in the direction of the flow. Pipes that are 3 inches, or larger, in diameter must have a clear space of 18 inches in front of the clean-out. On smaller pipes, the clear distance must be at least 12 inches. The sewer must be supported on a firm base of earth, sand, or gravel as it travels the length of the trench.

Water distribution systems

When you are ready to run water pipe below grade, there are a few rules you must obey. Pipe used for the water service must have a minimum working pressure of 160 pounds per square inch (psi) at 73.4 degrees Fahrenheit. If the water pressure in your area exceeds 160 psi, the pipe must be rated to handle the highest pressure available to the pipe. If the water service is a plastic pipe, it must terminate within 5 feet of its point of entry into the home. If the water service pipe is run in the same trench as the sewer, special installation procedures are required.

If you run the water service and the sewer in a common trench, you must keep the bottom of the water service pipe at least 12 inches above the top of the sewer pipe. The water service must be placed on a solid shelf of firm material to one side of the trench (Fig. 6-1). The sewer and the water service should be separated by undisturbed or compacted earth. The water service pipe must be buried at a depth that will protect it from freezing temperatures. The depth required will vary from state to state. Check with the code enforcement office for the proper depth in your area.

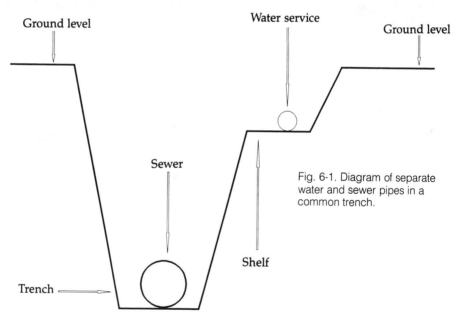

Fig. 6-1. Diagram of separate water and sewer pipes in a common trench.

Copper pipe run under concrete should be type "L" or type "K." If the copper pipe will come into direct contact with concrete, place it in a sleeve to protect the copper. Concrete may damage copper when the two materials come in direct contact. The pipe may also vibrate during use and wear a hole in itself as it rubs against rough concrete. All water pipe should be installed in a way that prevents abrasive surfaces from coming in contact with the pipe.

Tools and materials

The tools needed to install underground plumbing are minimal. You will need a round-point shovel, a pick, and possibly a tunneling bar for digging. A tape measure is mandatory equipment and a grade level is very helpful. You will need a torpedo level to set the grade on fittings. You will need a saw or some appropriate device to cut pipe. String is needed to keep your pipes inside the walls and at the proper distance from other walls. A hammer is useful for driving stakes in the ground and may be used with a chisel to cut through the foundation wall. A roller-type cutter will be needed if you are installing copper water pipe under the slab. This rounds out the list of tools most often needed to install underground plumbing.

The basics of working with the materials

One of the first pieces of knowledge you need is how to read a fitting. Fittings are read from bottom to top and then side to side. The readings are taken from the direction of flow. Refer to Fig. 6-2 for a clear understanding of how to read a fitting.

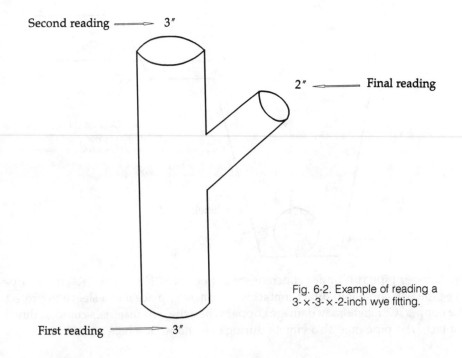

Fig. 6-2. Example of reading a 3- x -3- x -2-inch wye fitting.

In almost every circumstance, the pipe used for underground DWV plumbing will be schedule 40 plastic pipe. Working with this pipe is easy. The pipe may be cut with saws or roller-type cutters. Most plumbers cut the pipe with a saw. I use a hacksaw on pipe up to 4 inches in diameter. A handsaw, like carpenters use, is effective in cutting the pipe and is easier to use on pipe over 3 inches in diameter. There are saws made especially for cutting schedule 40 plastic pipe, but I don't like them. These saws are illustrated in Chapter 1. I have used them, but I prefer a hacksaw.

It is important to cut the pipe evenly. If the pipe is cut crooked, it will not seat completely in the fitting. This may cause the joint to leak. After you make a square cut on the pipe, clear the pipe of any burrs and rough pieces of plastic. You can usually do this with your hand, but the burrs and plastic might be sharp. It is safer to use a pocket knife or similar tool to remove the rough spots. The pipe and the interior of the fitting's hub must be clean and dry. Wipe these areas with a cloth, if necessary, to clean and dry the surface.

Apply an approved cleaning solution to the end of the pipe and the interior of the fitting hub. Next, apply an approved primer to the pipe and fitting hub. Now apply the solvent or glue to the pipe and hub. Insert the pipe into the hub until it is seated completely. Turn the pipe a quarter of a turn to assure a good seal. This type of pipe and glue makes a quick joint. If you make a mistake, you will have trouble removing the pipe from the fitting. If the glue sets for a long time, you will have to cut the fitting off the pipe to correct a mistake.

Many plumbers dry-fit their joints to confirm their measurements. Dry-fitting is putting the pipe and fittings together without glue to check alignment and measurements. I have done plumbing for a long time and don't dry-fit. I never did it in the early years, but many plumbers prefer to dry-fit each joint. This procedure makes the job go much slower and may cause its own problems. If you dry-fit the pipe to get an accurate feel for your measurements, the pipe must seat all the way into the fitting. When this is accomplished, it might be difficult to get the pipe out of the fitting.

While I never dry-fit, I do sometimes verify my measurements with an easier method. I place the fitting beside the pipe, with the hub positioned as it will be when glued. I then take a measurement to the center of my fitting to confirm the length of the pipe. By holding the pipe beside the fitting, instead of inserting the pipe, I don't have to fight to get the pipe out of the fitting. Once a joint is made with schedule 40 plastic pipe, it should not be moved for a minute or so. It doesn't take long for the glue to set up and then you may continue working without fear of breaking the joint.

When you install water pipe under concrete, avoid making joints in the pipe. The pipe should be installed in continuous lengths. If you must splice the pipe together, check with your local plumbing inspector for an approved method. Normally, the water pipe will be run in separate, continuous lengths of pipe. This part of the job does not require you to solder, crimp, glue, or clamp the pipe. All you must do is cut it.

Plastic pipe is usually cut with a saw, like a hacksaw. Copper pipe is cut with a roller cutter. Position the cutter on the pipe with the roller blade on the spot you wish to cut. Turn the handle until there is pressure between the rollers and the roller blade. Rotate the cutter around the pipe. After every two turns, tighten the handle to achieve more pressure. Continue this procedure until the pipe is cut.

Installing the groundworks

The placement of the groundworks is critical to the plumbing of your home. Before you install the underground plumbing, you must do some careful planning. The first step is to lay out the plumbing. Chapter 4 dealt with plumbing layouts in general. This section is targeted directly at layouts for underground plumbing.

Locating the water service and sewer locations

When the groundworks are installed, the sewer and water service may or may not be installed already. If they are installed already, your job is a little easier. The pipes will be stubbed into the foundation, ready to be connected. If these pipes are installed, you may start with them and run the underground plumbing. If the water service and sewer are not installed, you will have to locate the proper spot for them.

Refer to your blueprints for the location of the water and sewer entrance. If the pipe locations are not noted on the plans, use common sense. Determine where the water service and sewer will be coming to the house. If you are working with municipal water and sewer, the public works department of your town or city can help you. The public works department will tell you where the connections will be made to the mains and how deep they will be. If you will be connecting to a septic system and well, find the locations for these systems.

You need to consider two things when you pick the location for the water service and sewer. The first is the most convenient location inside the home for plumbing purposes. The second is an exit point that will allow a successful connection to the main sewer and water service outside the home. Your decision should give priority to the connection with the mains outside the home. You can adjust the interior plumbing to connect to the incoming pipes, but you may not be able to adjust the existing outside conditions.

Once you pick a location for the sewer and water service, you are ready to lay out the remainder of the groundworks. The locations for your underground pipes will be determined partially by the blueprints. The blueprints will show fixture locations. These fixture locations determine where you must have pipes. In addition to any grade-level plumbing, you will have to look at the plumbing on upper floors. Where there are plumbing fixtures above the level of the concrete floor, you will have to rough-in drain pipes now.

When you turn pipes up out of the concrete for drains and vents, the location of the pipes will be crucial since many will be inside of walls that have not been built. If your pipe placement is off by even 1 inch, the pipe may miss the wall location. If this happens, you will have a pipe sticking up through the floor in the wrong place. It may be in a hall or a bedroom, but the pipe will have to be moved. To move the pipe, you will have to break and repair the new concrete floor. To avoid these problems, make careful measurements and check all measurements twice.

When you have decided on all the pipes you will need for the underground plumbing, lay out the ditches. Normally, you will have to dig ditches for the pipes. The easiest way to mark the ditches is with lime or flour. Place these white substances on the ground, in the path of the ditch, so you dig the ditches accurately. The ditches will have to be graded to allow the proper pitch on your pipes. The standard pitch for household plumbing is 1/4 inch of fall for every 12 inches the pipe travels.

Installing the sewer

When all the planning and layouts are done, you are ready to start working. The best place to start is with the sewer. The sewer may leave the house under the footing or through the wall. The height of the exit point will be dictated by the depth of the connection at the main sewer. If the sewer is not already stubbed into the foundation, you will have to establish the proper depth for the sewer where it leaves the foundation. Sometimes, if you take the sewer out under the footing, the drain will be too low to make the final connection to the main. Before you tunnel under the footing or cut a hole in the foundation, determine the proper depth for the hole. This may be done by measuring the distance from the main sewer to the foundation. If the main connection is 60 feet away, the drain will drop 15 inches from the time it leaves the house until it reaches the final destination.

The sewer pipe should be covered by at least 12 inches of dirt when it leaves the foundation, so the main connection must be at least 27 inches deep. This depth is arrived at by taking the 12 inches needed for cover and adding it to the 15 inches of drop in the pipe's grade. It it best to allow a few extra inches to ensure a good connection point.

I will start the instructions assuming that the water service and sewer have not been run to the house before you start plumbing. With the depth of the hole for the sewer known, tunnel under the footing or cut a hole in the foundation wall. Install a sleeve for the sewer that is at least two pipe sizes larger than the sewer pipe. Install a cap on one end of the sewer pipe. Extend the capped end of the sewer pipe through the sleeve and about 5 feet beyond the foundation. You are now ready to install the groundworks for all the interior plumbing.

Installing DWV pipes

The first fitting to install on the building drain is a clean-out. The clean-out usually will be made with a wye and an eighth bend. The pipe that extends from the eighth bend must extend high enough for the clean-out to be accessible above the concrete floor. Many plumbers put this clean-out in a vertical stack that serves plumbing above the first floor. The clean-out may be stopped at the finished floor level, but it must be accessible above the concrete.

After you install the clean-out, go on about your business. As you run the pipes to the appropriate locations, support them on solid ground or an approved fill, like sand or gravel. Maintain an even grade on the pipes as they are installed. The minimum grade should be 1/4 of an inch for every 12 inches the pipe runs. Don't apply too much grade to the pipe. If the pipe falls too quickly, the drains may stop up when used. The fast grade will drain the pipes of water, but leave solids behind. These solids accumulate to form a stoppage in the pipe and fittings.

Support all the fittings installed in the system. If you install a 3- × -2 wye, the 2-inch portion will be higher above the ground than the 3-inch portion. Place dirt, sand, or gravel under these fittings to support them. Don't allow dirt, mud, or water to get in the fittings or on the ends of the pipe. A small piece of dirt is all it takes to cause a leak in the joint.

I said earlier that you might need string to lay out your wall locations so you know where to position the pipes for fixtures, vents, and stacks. Also, you should know what the finished floor level will be. The blueprints will show wall locations and the general contractor, concrete sub-contractor, or foreman will be able to give you the finished floor grade.

You must know the level of the finished floor so that the pipes will not get too high and wind up above the floor. Once you have the grade level for the finished floor, mark the level on the foundation. Drive stakes into the ground and stretch a string across the area where the plumbing is being installed. The string should be positioned at the same level as the finished floor. Use this string as a guide to monitor the height of your pipes.

Using the blueprints, mark wall locations with stakes in the ground. You may use a single string and keep the pipes turning up through the concrete to one side of the string, or you may use two strings to simulate the actual wall. Using two strings will allow you to position the pipes in the center of the wall. Refer to Fig. 6-3 for an example of these methods.

Plumbing measurements are generally given from the center of a drain or vent. If your water closet is supposed to be roughed in 12 inches off the back wall, the measurement is made from the edge of the wall to the center of the drain. Refer to Fig. 6-4 for an example of these measurement methods.

As you complete the underground DWV system, secure all the pipes so that they stay in the proper position. Be most concerned with the pipes that come up through the concrete and the pipes for bathtubs, showers, and floor drains. The

Fig. 6-3. Strings may be used to represent future wall locations.

Fig. 6-4. Measurements are made from the edge of a wall to the center of the drain or vent.

best way to secure these pipes is with stakes. The stakes should not be made of wood. In many regions, termites may come to the house to eat the wood under the floor. Use steel or copper stakes. The stakes should be on both sides of the pipe to ensure that it is not moved by other trades (Fig. 6-5).

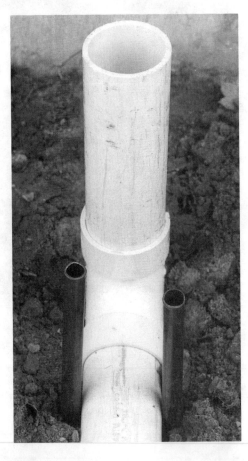

Fig. 6-5. Groundworks may be staked in place to avoid accidental displacement.

When all the DWV pipe is installed, cap all the pipes. You may use temporary plastic caps or rubber caps, but cap the pipes. You do not want foreign objects to get into the drains and clog them. Where you will install a tub or shower on the concrete floor, you will need a trap box (Fig. 6-6). This is just a box that keeps concrete away from the pipe and allows the installation of a trap when the bathing unit is set. Put a spacer cap over the turned up pipe for a toilet that will be set on the concrete floor (Fig. 6-7). The spacer will keep concrete away from the pipe so that you can install a closet flange when the time comes to set the toilet.

Fig. 6-6. A trap box is used to restrict concrete from blocking access to a tub or shower installation.

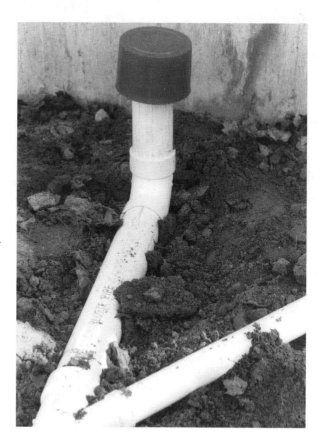

Fig. 6-7. A spacer cap prohibits concrete from causing an obstruction when you set a toilet.

Installing water pipe

Installing the underground water pipe is a breeze. These pipes should be installed without any joints under the concrete. Pick the place where you will be setting the water heater. Use this location to start all of your hot water pipes. If you are using copper pipe, use coiled, soft copper. It should be type "L" or type "K." Run a separate pipe for each fixture served, if the service is coming from below the concrete floor. All you have to do is lay loops of pipe onto the ground.

Make sure the pipe is not in contact with any rocks or rough surfaces. Put a copper stake in the ground to secure the pipe where it turns up to come out of the concrete. You may attach the water pipe to the stake with duct tape (Fig. 6-8).

Fig. 6-8. Underground water pipes may be secured in place with duct tape wrapped around copper stakes.

Cap both ends of the pipe to keep out foreign objects. You may run all of your cold water loops to the water heater location, but mark the pipes so you know which are hot and cold. A colored tape works well for marking pipes. The last step is to run a loop of pipe from the water heater to the water service. Support all the water pipe with a solid base and be sure the pipe is covered with a soft material. If stone is dumped on the water pipe, you may have leaks under the concrete later.

The wrap-up

This concludes the information you need to install underground plumbing for the average job. Installing the groundworks is not difficult, but you must make sure your measurements are accurate. After the groundworks are installed, check the measurements on all the pipes. It is much easier to change mistakes before the concrete is poured than it is afterwards.

7

Installing the drain-waste-vent system

THE INSTALLATION OF the drain-waste-vent (DWV) system intimidates many people. The job often is considered the most complex task when plumbing a house. The process is governed by many rules and regulations, but it need not intimidate you. When you learn to apply basic principles, the installation of the DWV system will go smoothly. Unless you are plumbing a complicated fixture, like an island sink, the basics of installing the DWV system are not overwhelming.

One key to an easy installation is good planning. If you look ahead and plan your system well, the installation will go together quickly. Observing standard rough-in dimensions will keep the pipes in the proper place. If you follow the instructions in this chapter, I believe you will be able to install your own DWV system.

Filling the toolbox

The tools used to install DWV piping are not complicated or numerous. You will need a tape measure, hammer, step ladder, torpedo level, and a saw for cutting pipe. Wood chisels are sometimes needed and a screwdriver and adjustable wrench might come in handy. The power tools you will need are a right-angle, 1/2-inch drill, and a reciprocating saw. An extension cord will be needed to run the power tools. A pencil and a lumber crayon finish the list of tools required to rough-in DWV pipes above grade.

Power tool safety

If you are an amateur, you may have to rent the right-angle drill and reciprocating saw. These tools are potentially dangerous and deserve your respect. When operating either of these tools, wear safety glasses (Fig. 7-1). Your eyes are

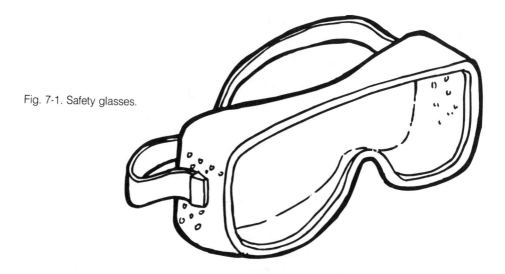

Fig. 7-1. Safety glasses.

irreplaceable and they can be damaged quickly doing plumbing work. Now let's look at specific safety procedures for the power tools.

The reciprocating saw is not unusually dangerous when used properly. Most of these saws are easy to operate and are turned on by squeezing a trigger. Some models have a button to lock the trigger in the "on" position. Do not use these trigger locks. When inexperienced people operate these saws, the blades often break. If you are wearing safety glasses, the broken blades should not present a problem.

When you use a reciprocating saw, do not put yourself in a position where you could be cut. I know this sounds ridiculous, but you would be surprised how many people put their hand in the path of the saw. Don't wear loose, baggy clothes. The saw blade might become entangled in the clothing and work its way to your skin. Inspect the electrical cord for broken or cracked insulation. If the insulation is worn or broken, don't use the saw. Never use any power tool while standing in water. While we are on the subject of water, never cut pipes filled with water with a power saw.

The right-angle drill is more dangerous than the reciprocating saw. This $1/2$-inch drill is powerful and can do nasty things to your body. I know from experience that these drills can get your respect fast. I still carry the scars from an incident with a right-angle drill. I have also had many helpers ignore my warnings about the drill and wind up in great pain.

If you are drilling through studs in a wall, keep your head away from the drill. If the drill hits a nail or a knot, it will kick upward. If your head is in the way, you will get a headache or lose a tooth. Don't hold the drill in such a way that your fingers might get caught between the handle and a piece of wood if the drill kicks. When you drill above your head, be very careful not to overextend yourself. Do not attempt to drill holes through the plywood roof sheathing for vent pipes. Cut these holes with the reciprocating saw. When the worm-driver of the bit gets through the thin plywood, the teeth come into contact with the wood and the drill will try to knock you out of the attic. Now that I have adequately terrified you with right-angle drills, let's move on.

Getting started

You have already learned how to design, size, and lay out the system. At this point, all that is left is to install the pipe. The physical act of putting the pipe and fittings together is easy. The most difficult aspects are code requirements and planning ahead. Failing to pay attention to either of these elements will result in an unsatisfactory job. Code requirements can be learned from reading the code book. Good planning can be learned from reading this book. With the combined reading, you should do fine on the DWV installation.

Mark all the holes and check the clearances for the holes. Be sure to check the area below your hole. You do not want to drill into electrical wires, steel, headers, or other drill-stopping obstacles. When everything looks good, cut the holes. This is where the installation begins. Cut the holes for the toilets, and for the tub and sink drains. These holes will show you exactly where you must run your pipes.

Hole sizes are important.   Some codes require you to keep the hole size to a minimum to reduce the effect of fire spreading through a home. If you drill over-sized holes, they might act as a chimney for fires. Choose a drill bit that is just slightly larger than the pipe you will be installing. In the case of 2-inch pipe, a standard drill bit size will be $2^9/_{16}$ inch.

When you cut a hole for a shower drain, the hole size will be determined by the size of the shower drain. Keep the hole as small as possible. Make it just large enough that the shower drain will fit in it. For a tub waste, the standard hole will be 15 inches long and 4 inches wide. This allows adequate space for the installation of the tub waste. In some jurisdictions, after the tub waste is installed, the hole must be covered with sheet metal to eliminate the risk of a draft during a house fire.

Running pipe

When the design is made and the holes are open, you are ready to run pipe. This is where you must exercise your knowledge of the plumbing code. Many rules pertain to the DWV system. If you purchase a code book, you will see page upon page of rules and regulations. Installing the pipes for the DWV system scares many people, plumbers included. The task seems overwhelming with all the rules to follow. If you don't lose your composure, the job is not all that difficult. By following a few rules and basic plumbing principles, you can install your own DWV system.

One of the most common mistakes made in the drainage piping is failing to remember to allow for the pipe's grade. Once you know the starting and ending points of the drains and vents, you can project the space needed for adequate grade. Generally, the grade is set at 1/4-inch-per-foot for drains and vents. Drains fall downward, toward the final destination. Vents pitch upward, toward the roof of the house. With a 12-foot piece of pipe, the low end will be 3 inches lower than the high end. When you are drilling through floor joists, most holes will be kept at least 1 1/2 inches above the bottom and below the top of the joist. If you follow this rule-of-thumb, you will have 4 1/2 inches to work with on a 2- × -8-inch joist.

A 2×8 has a planed width of 7 1/2 inches. From the 7 1/2 inches you deduct 3 inches for the top and bottom margin. This leaves you with 4 1/2 inches. What does all this mean to you? It means with a 2- × -8 joist system, you can run a 2-inch pipe for about 10 feet before you are in trouble. The pipe diameter uses 2 inches of the remaining 4 1/2 inches. This leaves 2 1/2 inches to manipulate for grade. At 1/4 of an inch per foot, you can run 10 feet with the 2-inch pipe. With a 1 1/2-inch pipe, you can go 12 feet. You will be restricted to a distance of about 6 feet with 3-inch pipe.

Under extreme conditions, you can run farther by drilling closer to the top or the bottom of the floor joist. Before drilling any closer than 1 inch to either edge, consult the carpenters. They will probably have to install headers or a small piece of steel to strengthen the joists. Many experienced plumbers run out of space for pipes. They don't look ahead and do the math before drilling the holes. After drilling several joists, they realize they cannot get to where they want to go. This results in a change in the layout and a bunch of joists with holes drilled in them that cannot be used. Plan your path methodically and you will not have these embarrassing problems.

Except for jurisdictions using a combination waste-and-vent code, you must provide a vent for every fixture. This does not mean every fixture must have an individual back-vent. Most codes allow the use of wet vents. Using wet vents will save you time and money.

Wet vents

You install a wet vent by using the drain of one fixture as a vent for another fixture. Toilets are often wet vented with a lavatory. To do this, you place a fitting within a prescribed distance from the toilet that serves as a drain for the lavatory. As the drain proceeds to the lavatory, it will turn into a dry vent after it extends above the trap arm. Refer to Fig. 7-2 for an example of a wet-vented fixture.

Dry vents

Many of the fixtures will be vented with dry vents. These are vents that do not receive the drainage discharge of a fixture. Since the pipes only carry air, they are called dry vents. There are many types of dry vents, including: common vents (Fig. 7-3), individual vents (Fig. 7-4), circuit vents (Fig. 7-5), vent stacks (Fig. 7-6), and relief vents (Fig. 7-7), to name a few. Don't let this myriad of vent

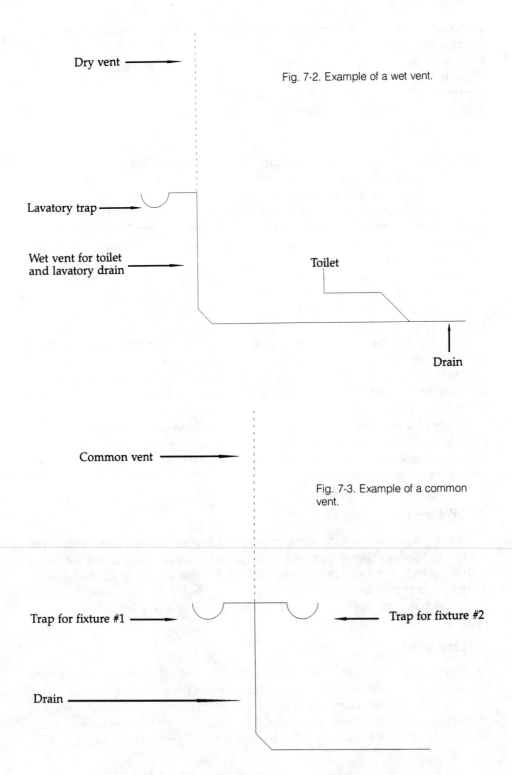

Dry vent

Fig. 7-2. Example of a wet vent.

Lavatory trap

Wet vent for toilet
and lavatory drain

Toilet

Drain

Common vent

Fig. 7-3. Example of a common
vent.

Trap for fixture #1

Trap for fixture #2

Drain

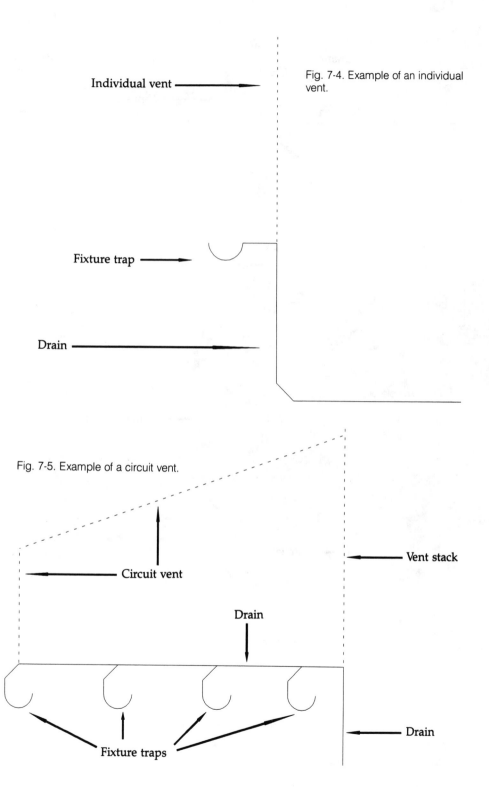

Individual vent

Fig. 7-4. Example of an individual vent.

Fixture trap

Drain

Fig. 7-5. Example of a circuit vent.

Vent stack

Circuit vent

Drain

Drain

Fixture traps

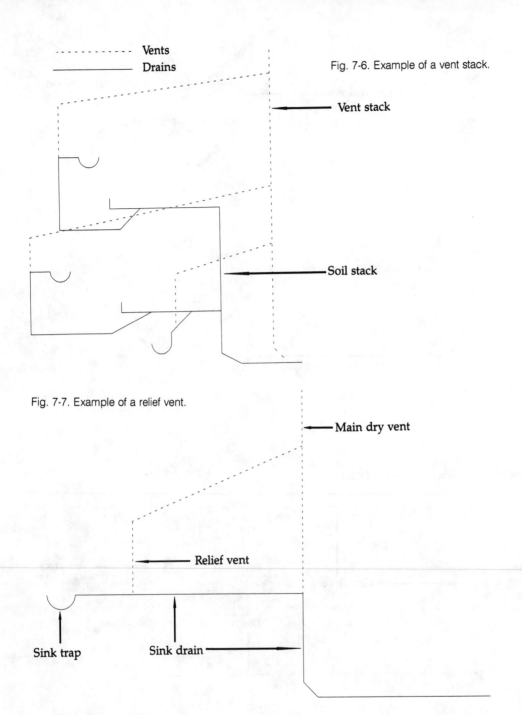

- - - - - - - Vents

——————— Drains

Fig. 7-6. Example of a vent stack.

Vent stack

Soil stack

Fig. 7-7. Example of a relief vent.

Main dry vent

Relief vent

Sink trap   Sink drain

types intimidate you. Check in the glossary at the back of this book for a definition of each type of vent, but don't dwell on the different types. When you plumb the average house, venting the drains does not have to be complicated.

By now, you may be starting to feel that installing the DWV system is beyond your capabiltities. If you are experiencing these feelings, set them aside. I am about to show you the easy way to plumb your DWV system.

Putting the system together

If you understand how to size the pipe for the DWV system, you are halfway home. There are only a few other rules that you must obey. The drains must have an even grade on them. A standard pitch for the pipes is 1/4 of an inch of fall for each foot the pipe travels. For your purposes, this is the only grade you will need to remember. Keep the grade consistent; don't allow great fluctuations in the fall of the pipe.

Drains should fall in the direction of the sewer. In other words, as your drain leaves the fixture, it should drop at a rate of a 1/4 of an inch per foot as it travels toward the building drain and sewer. All drainage fittings must be installed so that the throat of the fitting flows toward the main drain and sewer. Figure 7-8 shows you what I mean by the throat of the fitting.

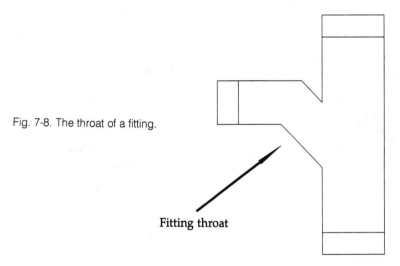

Fig. 7-8. The throat of a fitting.

Fitting throat

Vents should have the same standard grade, but they should be graded upward, towards the roof of the house. If you are installing a tee in a vent, the throat should flow upward. For simplicity, the fittings in the dry vents will be upside down when compared to drainage fittings. If you think about it, this all makes sense. Vents flow upward and drains flow downward.

If you don't understand wet vents, you may run all dry vents. This may cost a little more in material, but it might make the job easier for you to understand. Remember, every fixture needs a vent. How you vent the fixture is up to you, but as long as a legal vent is installed, you will be okay. You must install at least one 3-inch vent through the roof of the home. You may have more than one 3-inch vent, but you must have at least one.

After you have the mandatory 3-inch vent, most bathroom groups may be vented with a 2-inch vent. The majority of individual fixtures may be vented with a 1¹/₂-inch pipe. Most secondary vents may be tied into the main 3-inch vent before the main vent leaves the attic. In a standard application, the average house will have one 3-inch vent going through the roof and one 1¹/₂-inch vent that serves the kitchen sink and goes through the roof on its own.

If you want, you may vent each fixture with a separate vent and take all of them through the roof. This would be a waste of time and money, but you would be in compliance with the plumbing code. It is more logical to tie most of the smaller vents back into the main vent before it exits the roof.

Before you may change the direction of a vent, it must be at least 6 inches above the flood-level rim of the fixture it is serving. Figure 7-9 shows how this rule applies to your installation.

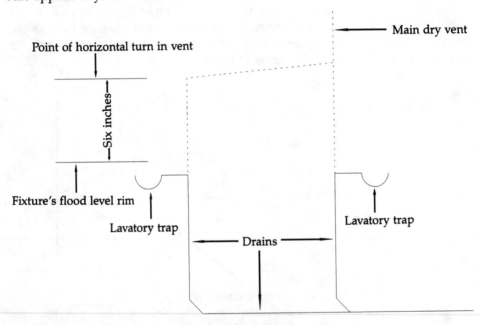

Fig. 7-9. Example of a vent tying into another vent.

When you start to install the drains, you must pay attention to pipe size, pitch, pipe support, and the fittings. The sizing charts in Chapter 3 will help you identify the proper pipe sizes. If you grade all of the pipes with the standard ¹/₄ of an inch per foot, you won't have a problem. Support all the horizontal drains at 4-foot intervals. The last thing to learn about is the use of various fittings.

When each fixture in the system is vented, use P-traps (Fig. 7-10) for tubs, showers, and washing machine drains. Where the pipes are rising vertically, use a sanitary tee (Fig. 7-11) to make the connection between the drain, vent, and trap arm. Use long-sweep quarter bends (Fig. 7-12) for horizontal direction changes. You may use short-turn quarter bends (Fig. 7-13) above grade when the

Fig. 7-10. PVC P-trap.

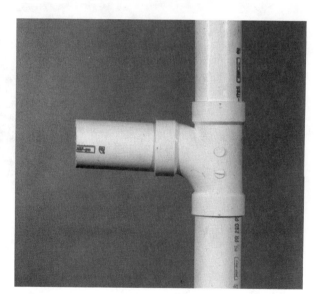

Fig. 7-11. PVC sanitary tee.

pipe changes from horizontal to vertical. But, if you want to keep it simple, use long-sweep quarter bends for all 90-degree turns.

Use wyes with eighth bends (Fig. 7-14) to create a stack or branch that changes from horizontal to vertical. Remember to install clean-outs near the base of each stack (Fig. 7-15) and at the end of horizontal runs (Fig. 7-16) when feasible. Keep turns in the piping to a minimum. The less turns the pipe makes now, the fewer problems you will have with drain stoppages later. Never tie a vent into

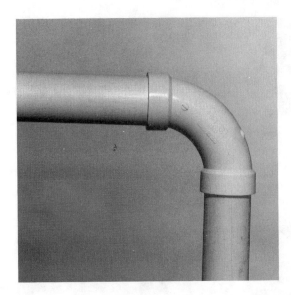

Fig. 7-12. PVC long-sweep quarter bend.

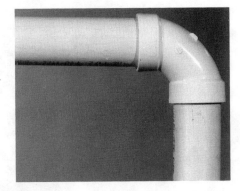

Fig. 7-13. PVC short-turn quarter bend.

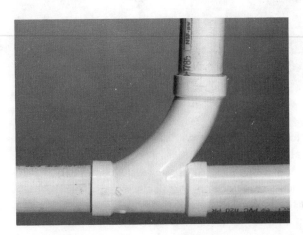

Fig. 7-14. PVC combination wye and eighth-bend fitting.

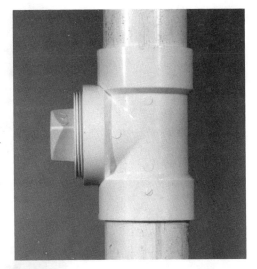

Fig. 7-15. PVC test tee and plug.

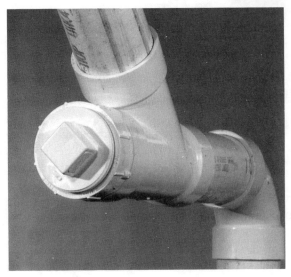

Fig. 7-16. Clean-out at the end of a horizontal pipe run.

a stack below a fixture if the stack receives that fixture's drainage. Where the pipe might be hit by a nail or drywall screw, install a metal plate to protect the pipe (Fig. 7-17). These nail plates are nailed or driven onto studs and floor joists to protect the pipe.

Keep the vents within the maximum distance allowed from the fixture. A $1^{1}/_{2}$-inch drain may not have a distance of more than 5 feet between the fixture's trap and the vent. For 2-inch pipe, the distance is 8 feet for a $1^{1}/_{2}$-inch trap and 6 feet for a 2-inch trap. Three-inch pipe will allow you to develop a maximum distance of 10 feet. With 4-inch pipe, you can go 12 feet. If the distance between the fixture and stack vent is greater than allowed, install a relief vent.

Fig. 7-17. Nail plate used to protect pipes.

A relief vent is a dry vent that comes off the horizontal drain as it goes to the fixture. The relief vent rises up at least 6 inches above the flood-level rim and ties back into the stack vent.

If you are faced with a sink in an island counter, you have some creative plumbing to do for the vent. Since the sink is in an island, there will be no walls to conceal a normal vent. Under these circumstances, you must employ an island vent (Fig. 7-18).

The vents that penetrate the roof must extend at least 12 inches above the roof. In some areas, the extension requirement is 2 feet. If the roof is used for any purpose other than weather protection, the vent must rise 7 feet above the roof. Be careful when you take vents through the roof near where windows, doors, or ventilating openings are present. Vents must be 10 feet away from these openings or 2 feet above them (Fig. 7-19).

When you plumb the drain for the washing machine, keep this in mind. The standpipe from the trap must be at least 18 inches but not more than 30 inches high (Fig. 7-20). The piping must be accessible for clearing stoppages.

If the fixture locations are not shown on the blueprints, you will have to pick the placement. There are some rules that apply to the distances beside and in front of fixtures. Let's talk about roughing-in the toilet.

Standard toilets will rough-in with the center of the drain 12 inches from the back wall (Fig. 7-21). The measurement of 12 inches is figured from the finished wall. If you are measuring from a stud wall, allow for the thickness of drywall or whatever else will be used to finish the wall. Most plumbers rough-in the toilet 12 1/2 inches from the finished wall. The extra 1/2 of an inch gives you a little

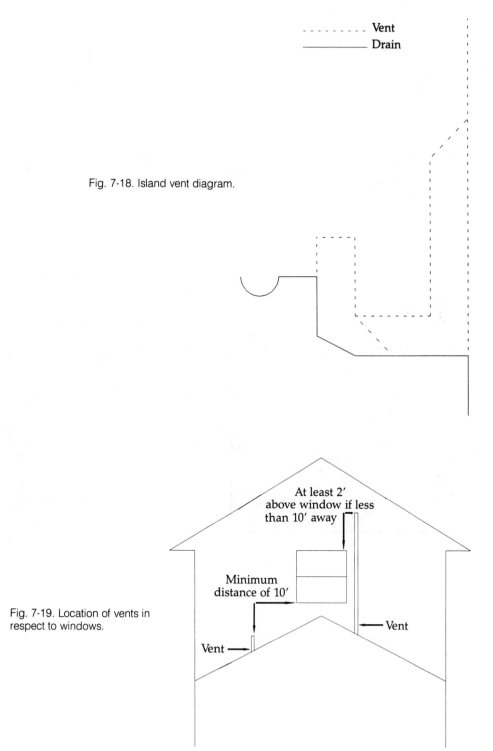

Vent ⎯ ⎯ ⎯
Drain ⎯⎯⎯

Fig. 7-18. Island vent diagram.

At least 2'
above window if less
than 10' away

Minimum
distance of 10'

Vent ⎯⎯

Vent ⎯⎯

Fig. 7-19. Location of vents in
respect to windows.

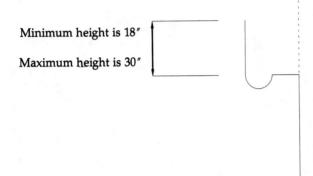

Minimum height is 18″

Maximum height is 30″

Fig. 7-20. Diagram of washing machine receptor rule.

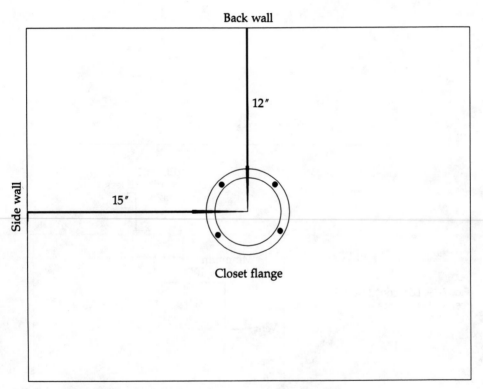

Back wall

12″

Side wall

15″

Closet flange

Fig. 7-21. Standard toilet rough-in dimensions.

breathing room if conditions are not exactly as you planned. From the front rim of the toilet, you must have a clear space of 18 inches between the toilet's rim and another fixture.

From the center of the toilet's drain, you must have 15 inches of clearance on both sides. This means you need a minimum width of 30 inches in the area in which you install a toilet. If you are plumbing a half-bath, the room must have a minimum width of 30 inches and a minimum depth of 5 feet. The same clearance measurements apply to bidets.

If you plan to install a tub waste using slip-nuts, you will have to have an access panel so that you can get to the tub waste. If you do not want to have an access panel, use a tub waste with solvent-weld joints. Many people object to access panels in their hall or bedroom. Avoid slip-nut connections and you can avoid access panels.

We have covered most of the rules you will have to obey when you put the DWV system together. Refer to your code book before doing the plumbing. If you have questions, talk with the plumbing inspector.

8

Installing the
water distribution
system

INSTALLING THE water distribution system for an average home is not complicated. Pipe sizing is simple and the rules governing the installation are not formidable. When you work with these small pipes, finding passable routes is simple. If you are able to install a DWV system, you can certainly install a water distribution system.

Chapter 2 informed you of the many types of pipes and fittings that may be used for a potable water system. Chapter 3 gave you examples of how to size the pipes in the water system. This chapter will teach you how to install the pipes and fittings for a working water distribution system.

Aside from how the connections are made, CPVC and copper systems will be installed using the same principles. Polybutylene systems may be installed along the same lines as copper or CPVC, but a manifold installation makes more sense. Well systems, pumps, and water conditioners are covered in Chapter 11. In this chapter, the instructions given will be based on jobs having a municipal water supply.

Since a majority of water systems are piped with rigid pipe, the bulk of these instructions will be based on the use of copper pipe. The same basic rules will apply to CPVC pipe. After the installation instructions for rigid pipe are complete, I will talk about polybutylene systems. Now, let's look at how you will plumb the water distribution system in your home.

Finding the starting point

The starting point for your potable water system will be where the water service enters the house. You will connect the main cold water pipe to the water service.

In most places, you will be required to install a back-flow preventer on the water service pipe. If the water service is polyethylene pipe, it must be converted to a pipe rated for hot and cold water applications. Polyethylene cannot handle the high temperature of hot water. The code requires you to use the same type of pipe for both the hot and cold water. Since polyethylene cannot be used for hot water, it may not be used as an interior water distribution pipe in a house that has hot water.

Convert the polyethylene pipe to the pipe of your choice with an insert adapter. The insert portion will fit inside the polyethylene pipe and be held in place by stainless steel clamps. The other end of the fitting will be capable of accepting the type of pipe you are using. In most cases, the adapter will be threaded on the conversion end. The threads allow you to mate any type of female adapter to the insert-by-male adapter. See Fig. 8-1 for an example of converting polyethylene pipe to copper.

Fig. 8-1. A male insert adapter and a female copper adapter mates copper pipe to polyethylene pipe.

Running the pipe

Once the conversion is made, install a gate valve on the water distribution pipe (Fig. 8-2). When required, install a back-flow preventer after the gate valve. It is a good idea to install another gate valve after the back-flow preventer. By isolating the back-flow preventer between the two gate valves, you can cut the water off on both sides of the back-flow preventer. This option will be appreciated if you must repair or replace the back-flow preventer.

As you bring the pipe out of the second gate valve, consider installing a sillcock near the rising water main. If the location is suitable, you will use a minimum amount of pipe for the sillcock. The main water distribution pipe for most houses will be a 3/4-inch pipe. If you have already designed the water distribution layout, install the pipe accordingly. If you haven't made a design, do it now.

The 3/4-inch pipe should go to the inlet of the water heater undiminished in size. You may choose to branch off the main to supply water to fixtures as the main goes to the water heater. This is acceptable and economical. Look back to Chapters 3 and 4 to refresh your memory on typical layouts.

The water pipe will normally be installed in the ceiling joists so they will be hidden when the ceiling is installed. When possible, keep the pipe at least 2

Fig. 8-2. Gate valve.

inches from the top or the bottom of the joists. This reduces the risk to the pipes of being punctured by nails or screws. If the pipe passes through a stud or joist where it might be punctured, protect it with a nail plate.

Water pipe should be hung and secured at 6-foot intervals. Avoid placing water pipes in outside walls and areas that will not be heated, such as garages and attics. The main delivery pipe for the hot water will originate at the water heater. When piping the water heater, the cold water pipe should be equipped with a gate valve before it enters the water heater. Many places require the installation of a vacuum breaker at the water heater. These devices prohibit the back-siphonage of the water heater's contents into the cold water pipes (Fig. 8-3).

Do not put a cut-off valve on the hot water pipe leaving the water heater. If a valve was installed on the hot water side, it might become a safety hazard. Should the valve close, the water heater might build excessive pressure. The combination of a closed valve and a faulty relief valve might result in an exploding water heater. Refer to Fig. 8-4 for a standard water heater installation.

Normally, you won't install valves on the fixtures during rough-in. Most fixtures get their valves during the finish plumbing stage. If you install the sillcocks or hose bibs during the rough-in, install stop-and-waste valves in the pipe that supplies water to the sillcock or hose bib (Fig. 8-5). Stop-and-waste valves have an arrow on the side of the valve body. Install the valve so the arrow is pointing in the direction that the water is flowing to the fixture.

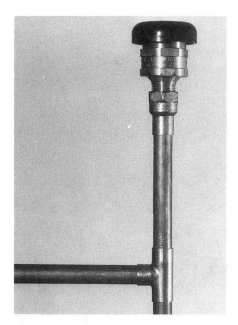

Fig. 8-3. Vacuum breaker
screwed into a female adapter.

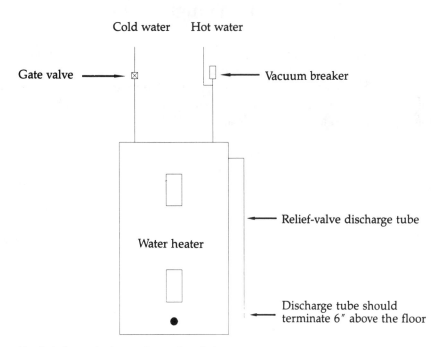

Cold water   Hot water

Gate valve ──────▶ ⊠   ▯ ◀── Vacuum breaker

Water heater

◀──── Relief-valve discharge tube

◀──── Discharge tube should
terminate 6″ above the floor

Fig. 8-4. A standard water heater installation.

Fig. 8-5. A stop-and-waste valve.

Rule-of-thumb rough-in numbers

Refer to a rough-in book to establish the proper location for the water pipes that serve fixtures. The examples given here are common rough-in numbers, but the fixtures may require a different rough-in. The water supplies for sinks and lavatories are usually placed 21 inches above the sub-floor. Kitchen water pipes are normally set 8 inches apart, with the center point being the center of the sink. Lavatory supplies are set 4 inches apart, with the center being the center of the lavatory bowl.

Most toilet supplies will rough-in at 6 inches above the sub-floor and 6 inches to the left of the closet flange as you face the back wall. Shower heads are routinely placed 6½ feet above the sub-floor. They should be centered above the bathing unit. Tub spouts are centered over the tub's drain and rough-in 4 inches above the tub's flood-level rim. The faucet for a bathtub is commonly set 12 inches above the tub and centered on the tub's drain. Shower faucets are generally set 4 feet above the sub-floor and centered in the shower wall. If you go to your supplier and ask for a rough-in book, the supplier should be able to give you exact information for roughing-in the fixtures.

Secure all pipes near the rough-in for fixtures. When you rough-in for a shower head, use a wing ell. These ells have ½-inch, female threads to accept the threads of a shower arm. The wing ell has an ear on each side that allows you to screw the ell to a piece of backing in the wall. Securing the ell will hold it in place for the later installation of the shower arm. Where necessary, install backing in the walls to which you can secure all water pipes. It is very important that you have all pipes firmly secured. Refer to Fig. 8-6 for an example of how to use a wing ell. Figures 8-7, 8-8, and 8-9 show other methods for securing the water pipes.

Fig. 8-6. A mounted wing ell.

Fig. 8-7. A copper mounting clip securing a horizontal copper pipe.

Fig. 8-8. Copper mounting clips securing vertical copper pipes.

Fig. 8-9. Copper wire hanger supporting a horizontal copper pipe.

If you are soldering joints, remember to open all valves and faucets before you solder them. If you don't, the washers and internal parts may be damaged by the heat. Consider installing air chambers to reduce the risk of noisy pipes later. The air chamber may be made from the same type of pipe you are using in the house. The air chamber should be at least one pipe size larger than the pipe it is serving. A standard air chamber will be about 12 inches tall. Refer to Fig. 8-10 for an example of how you might install an air chamber.

The installation of cut-off valves in the pipes feeding the bathtub or shower is optional. Most codes do not require the installation of these valves, but they come in handy when you have to work on the tub or shower or faucets in later years. Stop-and-waste valves are typically used for this application. Be sure to install the valves where you will have access to them after the house is finished. Avoid extremely long, straight runs of water pipe. If you are running the pipe a long way, install offsets in it. The offsets reduce the risk of an annoying water hammer later.

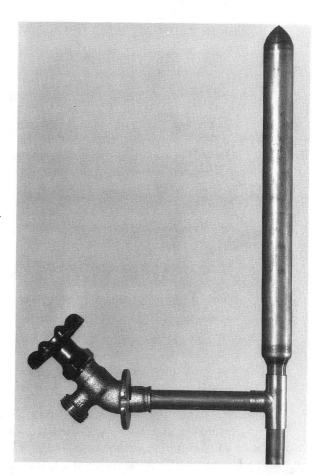

Fig. 8-10. Sillcock with an air chamber installed.

Plumbing a manifold system with polybutylene

The rules for plumbing the water distribution system are the same for all the approved materials. But, when running polybutylene, it often makes more sense to use a manifold system than to run a main with many branches. One of the most desirable effects of this type of installation is the lack of concealed joints. All the concealed piping is run in solid lengths without fittings. Refer to Figs. 8-11 and 8-12 to see how a polybutylene manifold might be set up.

To plumb a manifold system, run your water main to the manifold. The manifold may be purchased from a plumbing supplier. It will come with cut-off valves for the hot and cold water pipes. Your cold water comes into one end of the manifold and goes out the other to the water heater. The hot water comes into the manifold from the water heater. The manifold is divided into hot and cold sections. The pipe from each fixture connects to the manifold. When there is a demand for water at a fixture, it is satisified with the water from the manifold. Once you have both the hot and cold supply to the manifold, all you have to do is run individual pipes to each fixture.

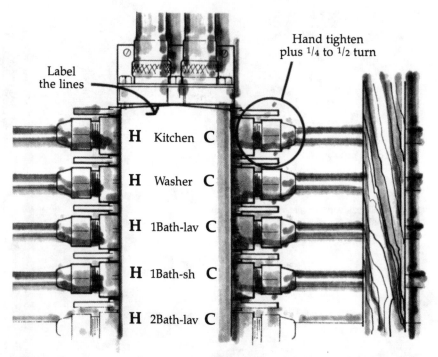

Fig. 8-11. Detail of a manifold used with a polybutylene piping system for water distribution. Vanguard Plastics.

You may snake the polybutylene pipe through the house in much the same way as you would electrical wires. Installing the water distribution system in this way saves many hours of labor. You also eliminate concealed joints, and you may cut off any fixture with the valves on the manifold. This type of system is efficient, economical, and the way of the future.

Like CPVC, polybutylene does not require any soldering skills or equipment. The pipe slides over a ridged insert fitting and is held in place with a special clamp. The clamps are solid metal and are installed with a special crimping tool. Many rental centers rent crimping tools by the day. Figure 8-13 shows the crimping tool used to make polybutylene joints.

Pipe contaminants

After water pipes are first installed, they need to be flushed. There will be many impurities in the pipes during the first uses of the plumbing. Before you consider the job done, clear the pipes of installation impurities. The types of impurities will depend somewhat on the type of pipe used for the water distribution system.

When you are ready to turn the water on for the first time, remove the aerators from the faucets. The objects coming through the pipes will clog the aerator

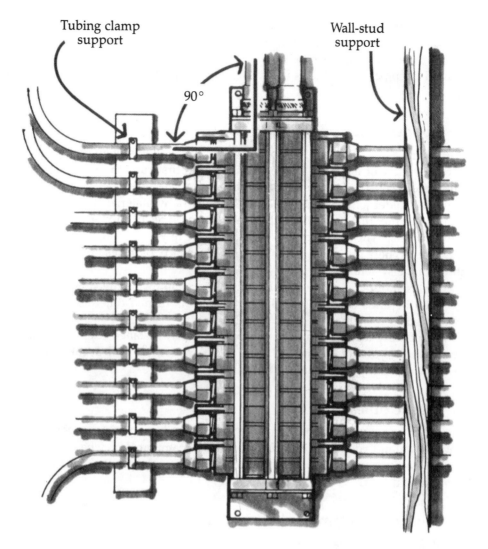

Tubing clamp
support

Wall-stud
support

90°

Fig. 8-12. Overview of a manifold used with a polybutylene piping system for water distribution. Vanguard Plastics.

Fig. 8-13. Crimping tool for polybutylene pipe.

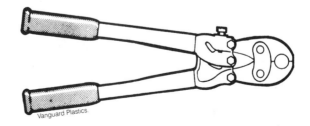

Vanguard Plastics.

Installing the water distribution system 105

screens. Remove the shower heads from the shower arms during the flushing. If you don't, the shower heads may become clogged with foreign particles. Don't connect the washing machine's hoses to the boiler drains until after you have cleared the pipes. The objects in the water will foul the screens in the washer's hoses.

Copper pipes

When you use copper pipes for the water distribution system, you might discover a variety of materials in the water when it first runs through the system. The water may contain green flakes. These green particles are the flux used to make the solder joints. If you see a thin, brown filmy material come out of your faucet, it is flux that has solidified. Small, sharp pieces of copper may come out of the faucets. When the copper pipe is cut, these tiny burrs sometimes remain in the pipe. When the water pressure hits the sharp particles, they are forced through the system.

Dirt is another possible contaminant. When you install the system, dirt may get into the ends of the pipe. Until the water is turned on, this dirt may go unnoticed.

All of these materials might be present in newly installed copper pipes. It is important for your health and the condition of the faucets to clear the water lines before you use the water for potable purposes.

Plastic pipes

When you use plastic pipe for the water distribution system, you will not have to contend with leftover flux, but other contaminants may be present. The curled pieces of plastic that occur when the pipe is cut might be found when the pipes are flushed.

Flushing contaminants from the water pipes

Before you use the new water system, clear it of contaminants. To flush the system, all you have to do is run water through all the valves and faucets. It is best if you start by running water through outside faucets and boiler drains. This will remove many of the larger objects that may foul the working parts of the faucets. Ultimately, you should run water through all of the valves and fixtures before you consider the system to be in working order.

Run the water through each valve and faucet for several minutes. When the water has run for awhile, cut it off. Allow the system to sit idle for a few minutes, and then repeat the cleansing process. When you allow the system to rest between flushings, you remove more of the particles in the water. When you feel the pipes are flushed, reinstall the aerators and shower heads. Now, turn on the water to each fixture again. If you get good pressure and an even stream of water, the pipes should be fine. If the pressure is low or the water sprays from the aerator, you still have some work to do.

When the aerators spray, it usually means they are partially blocked with sediment. If the pressure is low, you most likely have a clogged aerator. Remove the aerator and inspect the plastic disc and screen wire. If the holes in these devices are plugged, run the water for several minutes without the aerator on the faucet. You can usually clean the aerators and return them to a usable condition. If they are plugged beyond repair, you will have to replace them.

In closing

By now you should have a good idea of how to install the water piping. With the information in this chapter, you can make an informed decision on the type of pipe you want to use in the system. By combining this chapter with the information in Chapters 3 and 4, you should be able to install your own water pipe. Talk with your local plumbing inspector and review the local code requirements before you plumb the job. Confirming the regulations and requirements before you install the plumbing will save you money and headaches.

9

Code compliance and the rough-in inspection

AFTER YOU HAVE ALL of the plumbing roughed-in, you must test it and prepare it for inspection. The local plumbing inspector will pay you a visit and check your work. When the inspector arrives, it is important to have everything in order. For your own peace of mind, you will want to be certain there are no leaks in the system, especially ones that may be concealed later.

The intensity of the plumbing inspection will vary in different jurisdictions. I have worked in all types of environments and have experienced a wide range of inspection rituals. Some inspections have been very thorough and strict, others have been little more than a casual stroll through the job. To be safe, it is always best to have all your work in compliance with the code. This is not only best because it avoids failed inspections, it also protects you. The plumbing code was made to protect the public's health. If you adhere to the rules in the plumbing code book, your health and safety will be better served.

Testing the sewer

When you install a new sewer, the installation must be inspected by the plumbing inspector. When the inspector checks your work, he may look for many things. He will check to see that the pipe is of a proper size and that the grade on the pipe complies with the plumbing code. He may inspect the trench and how the pipe is supported. The distance between clean-outs may be measured and he will observe the fittings used to change directions in the pipe. The inspector will look to see that the water service is not too close to the sewer. When he is satisfied with these items, he will check for leaks in the piping. It will be your job to prepare the sewer for the leak inspection. The two most common ways of doing this are with water or air pressure.

Plugging the end of the sewer

Some locations will allow you to test the sewer up to the point of connection to the main sewer without making the final connection. This makes the test procedure easy. You simply cap the end of the pipe with a rubber or plastic cap. When the test is over, you remove the cap and connect the sewer to the main sewer. Normally, the connection will be made in the presence of the inspector.

Other jurisdictions require the final connection to be made prior to the test. This complicates the matter of plugging the sewer for the test. If you must connect to the main sewer before your test, you will have to install a test tee. The test tee is a flat tee, threaded for a clean-out plug. The test tee should be installed as close as possible to the final connection between your sewer and the main sewer. The threads should point upward, so you can get to them easily.

When you test with a test tee, you will need a test ball. The test ball is a heavy-duty, ridged balloon. You insert the deflated test ball through the test tee and into your sewer. Then, you use an air hose to inflate the test ball. Once inflated, the test ball plugs the sewer for the test.

There is an important safety factor that must be observed with test balls. If you decide to test the sewer with air, you have nothing to worry about, but if you test with water, be careful. When you have finished your test, the test ball must be removed. The removal is accomplished by letting the air out of the test ball. This simple act can result in disaster if you are testing with water. A 4-inch sewer with a 10-foot head of water applies substantial pressure to the test ball. The wrong removal process may result in a broken wrist, or worse.

The test ball will be equipped with a large ring on the end of a chain. This ring prohibits the loss of the test ball into the pipe; it also provides a means for removing the test ball. Never put your hand into the test tee to let air out of the ball. If your hand is in the test tee when the air pressure is released, the force of the water behind the test ball will do nasty things to your body parts.

Always use a long screwdriver, or similar tool, to release the air pressure from the test ball. Hold the large ring and create tension between the ring and the ball. Use the long screwdriver to let the air out of the balloon. As the test ball begins to deflate, pull it out of the pipe with the large ring and chain. This procedure requires some experience if you expect to stay dry. Unless you are accustomed to removing test balls, you will get wet. More importantly, do not put your hands into the test tee to remove the test ball.

Testing with air

If you use air pressure, you will have to cap off all open sewer outlets and clean-outs. You will connect an adaptor to one of the clean-outs so that air can be pumped into the pipe. This adapter will be a series of threaded, reducing bushings.

When the clean-out opening is reduced to a smaller size, you mount the test rig with male adapters or nipples. The test rig will have a tee in it. The pipe from the bottom of the tee will connect to the clean-out opening. The center outlet of the tee will hold the air valve. The top outlet will be fitted with an air-pressure gauge. Refer to Fig. 9-1 for a drawing of how the test rig will look.

Fig. 9-1. Test rig used to test DWV system.

Once the test rig is hooked up to the drainage piping, you will need to fill the pipe with air. You may use an air compressor or a tank holding pressurized air. The pipe must maintain a pressure of 5 pounds per square inch (psi), without leaking, for 15 minutes. I have had some inspectors wait the entire 15 minutes to see if the pipe leaked. If you find a slow leak in your preliminary test, correct it before the official inspection.

If you know you have a leak in the pipe, but can't find it, get some water and dish detergent. Mix the dish detergent with the water to create a soapy solution. Wipe the solution around the pipe joints. The one that is leaking will blow bubbles in the soapy solution. Once the leak is found, release the air pressure from the piping. Clean the pipe where it meets the fitting and apply a thick coating of glue where the pipe and fitting join. Let the glue sit for several minutes. After the joint has solidified, test the piping again. Hopefully, you will have solved the problem. If the pipe still leaks, cut out the leaking fitting and replace it with a new fitting and couplings.

Testing with water

When you test the sewer with water, the test procedure is different. You must cap the end of the sewer and fill it with water. You must have a pipe that extends at least 10 feet above the sewer. This pipe is usually a clean-out, but it must rise 10 feet above the sewer. When you fill the sewer with water, continue filling until the water is ready to run out of the 10-foot riser.

The inspector will watch the level of the water in the riser. If it drops, he will look for a leak. Smart inspectors will make you release the water from the sewer while they watch. This ensures that you did, in fact, have the sewer filled with water. Smart inspectors know how to outwit smart, but cheating plumbers. A cheating plumber will install a plastic test cap on the end of the 10-foot riser before he puts it into the fitting at the sewer. This allows the plumber to fill the riser full of water without filling the sewer. It saves time and effort for the plumber, unless he gets caught.

If you are testing with water and have a leak, drain the piping of all water. Clean and dry the pipe where it meets the leaking fitting. Apply fresh glue and wait for the glue to set-up. Refill the pipe with water and check your patch. If the leak persists, cut out the fitting and install a replacement. Be sure to keep the pipe and fitting free of water until the glue is dry.

The groundwork inspection

Preparing the groundwork for inspection will require attention to several details. As far as the actual test, the procedure is about the same as the test for a sewer. You may test with air or water. There must be a pipe standing 10 feet above the groundworks if you test with water. Most plumbers use a stack or a toilet riser for the test pipe. All the other pipes should be capped. You may cap the building drain before it is connected to the sewer. The methods used for testing the sewer may be used with groundworks. Provide a 10-foot head of water or 5 psi of air pressure.

In addition to the pressure test, there are many other little things to do. Make sure all of the pipes are evenly supported on a firm base. Maintain a consistent grade on all the pipes. Don't forget to place a sleeve around the building drain where it leaves the foundation. Don't backfill the pipe trenches with rocks. If you have water pipes turning up through the concrete floor, install sleeves to protect the pipes from the concrete. Install clean-outs as required by your local code. Make sure all of the underground pipes and traps are the proper size. In most codes, the minimum pipe size for drainage pipes under concrete is 2 inches.

Do not use short-turn quarter bends for any reason under the slab except as a a closet bend. It is acceptable to use a short-turn quarter bend where you go from horizontal to vertical for a closet flange. Make the job neat; inspectors are impressed with quality workmanship. With a little preparation, your inspection will go smoothly.

Testing the water service system

Normally, the water service is installed as a single length of pipe. Since there are no joints in the single length of pipe, you might think there is no need to test the pipe for leaks. The inspector will want to see the pipe and the installation, however, so don't cover the pipe before it is inspected.

Typically, the rough-in inspection for both the water pipe and the DWV system will be done at the same time. I am breaking the inspection into two categories to make the instructions easier to understand.

You can test the water pipes with air or water. Most professionals use air. It is easier to fix leaks if the pipes don't have water in them. Just as a note, soapy water will work to find leaks in the water pipes while there is air pressure on the pipes. There is a big difference between the amount of pressure required to test the water distribution system and the DWV system. The potable water system must be tested at a pressure equal to the working pressure of the system. This usually means a test with a pressure between 60 and 80 psi. Most professionals test their water pipes at 100 psi.

It is in your best interest to find any likely leaks at the rough-in stage. If the pipes spring leaks after the walls and ceilings are installed, you will be looking at some steep repair expenses when you have to cut into the new walls or ceilings. The test rig for air-testing the water pipe is about the same as the one used for DWV systems. The connection size is smaller, but the equipment is the same.

If you choose to test the water pipes with water, you will probably have to pump water into the pipes with a hydrostatic pump. If you have an operable well system or municipal connection, you may simply turn on the water. Obviously, you must cap all the pipes before turning on the water. Faucets that have a tub spout and a shower head should have the diverter activated to allow water or air to reach the shower-head riser. If you have to pump water into the pipes, you may rent a hydrostatic pump.

Most hydrostatic pumps are easy to use and connect to a standard garden-hose. You can connect the pump to a sillcock or a boiler drain at the washing machine hookup. The second hose on the pump will be placed in a bucket of water. Prime the pump by pouring water into the hose connected to the pump. As you move the handle up and down, the pump will fill the water pipes with water. As the water compresses the air already in the pipes, you build pressure. Watch your pressure gauge and don't allow it to go over 120 psi. One hundred pounds of pressure is plenty for household water pipes.

When you reach the target pressure, observe the gauge for several minutes. If the needle drops, look for the leak. When you find the leak, you will have to drain the water from the pipes until it is below the point of the leak. When the pipe is dry, correct the leak. In addition to the pressure test, make sure all other code requirements have been met in the water distribution system.

If you are forced to put water pipes in an unheated area, they must be insulated. There are many approved types of pipe insulation that you may use (Fig. 9-2). All of the water pipe should be supported at 6-foot intervals. All of the pipes should be secured to backing in the walls to eliminate vibration in the pipes. Pipe sizing must meet the minimum standards. You must use approved materials and make approved connections. The pipes must be protected from unintentional damage by other trades. An example of this protection is the use of nail plates. Refer to your local code book for exact specifications and be sure your installation is in compliance before the inspector arrives.

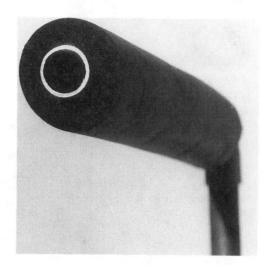

Fig. 9-2. Foam pipe insulation installed on a copper pipe.

Testing the DWV system

This is the inspection people fear more than any other. There is no reason to be concerned about the DWV inspection if you understand and follow the code and standard plumbing principles. There are a number of rules pertaining to DWV systems, but I have showed you how to simplify the rules. The DWV inspection will pass without a problem if you follow my advice and the rules in your code book.

You may test the DWV system with air or water. If you decide to test with air, you must cap all of the open pipes, including the vents on the roof. If you test with water, there is no need to cap the vent terminals. You may use plastic test caps (Fig. 9-3) or rubber, reusable caps (Fig. 9-4). To block the opening in a closet flange, you may use a test ball (Fig. 9-5) or an expansion plug (Fig. 9-6).

When you test with water, you must fill the system to the top of the pipe at the vent terminal. Some areas will allow a little grace in this situation. They allow

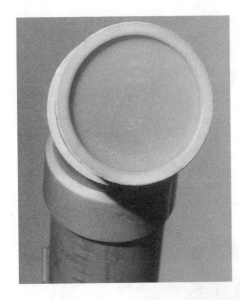

Fig. 9-3. Plastic test cap used during the testing of a DWV system.

Fig. 9-4. Rubber test cap used during the testing of a DWV system.

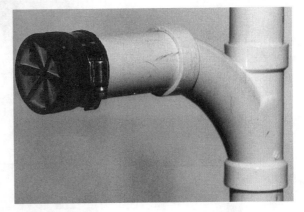

Fig. 9-5. Test ball to seal the opening of a closet flange during a test of the DWV system.

Fig. 9-6. Expansion plug to seal the opening of a closet flange during a test of the DWV system.

you to fill the system to the overflow level of the bathtub on the highest level. Their primary concern is finding leaks in the drainage. While you may be allowed to test only to the tub, plan on testing all the way through the roof.

If you test with air, the same ratings and rules apply as those for sewers and groundworks. If you test with air or water to the vent terminals, you may not install bathtub and shower traps before the test. If you connect the traps, you will have to block the pipes with test balls. Standard procedure calls for capping the pipes just below the tub or shower. After the test, you install the traps.

Test tees are often used as clean-outs in the vertical stacks of the DWV system (Fig. 9-7). Their installation allows you to use test balls to block the drainage pipes for the inspection (Fig. 9-8). Again, remember not to put your hand into

Fig. 9-7. Test tee in the base of a stack.

Fig. 9-8. Test ball to block a stack during a test of the DWV system.

Fig. 9-9. Test tee being used as a clean-out to allow the drain to be snaked.

the test tee to release pressure from the test ball. Test tees also allow you to run a snake down the pipe to clear any blockage occurring during the installation of the plumbing (Fig. 9-9).

What do you have to look out for beyond the pressure test? Well, the inspector may go over the DWV system with a fine-toothed comb. Make sure all the horizontal pipes are supported at 4-foot intervals. Maintain a constant grade on all the drains and vents. Be mindful of the fittings used to change directions in the pipe. Use approved materials in the appropriate sizes. Don't butcher floor joists and other structural members. Some codes require you to take care not to create a fire hazard with oversized holes. In these areas, keep your holes at a minimum size and seal gaps around oversized holes and cuts.

Be sure the vents extend the proper height above the roof. If the vents are near a window, door, or ventilating opening, install them 10 feet from these objects or extend them 2 feet above the openings. When necessary, install nail plates to protect the pipes.

Install the proper clean-outs and traps. Watch the height of the standpipe for the washing machine. It must be between 18 and 30 inches high. All of the vents must extend 6 inches above the flood-level rim of the fixture before they may be offset. Follow the guidelines in this book and rely on your local plumbing code book to verify your installation.

Dealing with the code officer

Code officers are people. That means they are individuals and largely unpredictable. As a plumber, the success of your installation depends on the code officer. If the plumbing inspector tells you to change your work, the odds are high you will have to change your installation. Even if you are right and the inspector is wrong, it is often best to make the suggested changes. Fighting the system may cause more trouble than it is worth.

As a young master plumber, I decided to set the record straight with a plumbing inspector who was incorrect in his decision. One of my jobs had been rejected on a bogus decision. Being young and an enthusiastic master plumber, I challenged the plumbing inspector. I knew he was wrong and I could document the facts in the code book. When the inspector disputed my claims, I went over his head. To make this long story short, I won the battle, but lost the war.

I was right in my position on the issue. The inspector's supervisor overturned the decision in my favor. I basked in the thrill of victory for awhile, then I paid the price. It seemed like all my inspections after this incident were very strict, not only from the corrected inspector, but from his colleagues as well. It took nearly a year for my inspections to get back to normal. I learned quickly it is not in the best interest of a professional plumber to flex his muscle against the inspectors.

You will do well to heed my advice and learn from my experience. Even if you know you are right, it is often easier to comply with the inspector's request than it is to challenge it. It may go against your grain, as it does mine, to give in, but there are times when silence is golden. Your interest is in getting an approved inspection in the least amount of time. That goal is best achieved by working with the code officers.

If you are a full-blooded maverick, as I was, let me tell you something about the code book. You can preach chapter and verse, and still be wrong. There is a universal clause inserted in the code that leaves the interpretation of the code book to the local authority. This single sentence removes any chance you might have of winning a knock-down dispute. Believe me, it is easier to make minor changes in the plumbing, than to go up against the local authorities.

It has been my experience that most code officers are reputable and willing to work with you. If you do not have an attitude problem, the inspectors will help you. When you go in with a chip on your shoulder, it is a good chance the inspector will knock it off. On the other hand, if you are humble, most inspectors will go out of their way to help you. Diplomacy is the key to a hassle-free inspection.

Avoiding problems with the plumbing inspector

After learning my hard lesson as a hot-shot master plumber, I learned how to successfully work with unknown inspectors. Any time I was hired to work in a new jurisdiction, I went to see the plumbing inspector before starting the job. I highly recommend you follow this same advice.

If you are not an established, local plumber, the inspector may take a longer look at your work. I have found a system that always seems to work. When I plan to work in a new venue, I meet with the plumbing inspector and ask for advice. Most inspectors are proud to expound on their knowledge by giving you the ins and outs of complying with their code.

If you go to the inspector and tell him that you will be working in his town, his ears will perk up. Follow your introduction with a request for any special conditions the local code requires. The code book is only a starting point for compliance. Remember that the final decision is left to the opinion of the local authority. If you ask the inspector for guidance, you are almost assured of a fair inspection. Don't make this visit as a mere gesture. Pay attention to the advice you are given and abide by it. There is no reason to make enemies by being stubborn. I learned this lesson the hard way. There is no reason you should.

The final word

When it comes to passing your plumbing inspection, the code enforcement officer has the final word. I cannot stress to you enough to develop a good working relationship with your inspector. I know from experience that life can be miserable if an inspector uses every available option to complicate your job. You are at enough of a disadvantage if you are not a master plumber; don't expand the threshold of pain by being a know-it-all.

10

Setting fixtures

TRIM-OUT PLUMBING can appear deceivingly simple. When you watch an experienced plumber set fixtures, the job looks like work almost anyone can do. Well, almost anyone can do the job, but for most, it will not be as easy as it looks. Without knowing the secrets for successfully setting fixtures, your attempts might be costly and frustrating. There is an art to installing plumbing fixtures. This chapter explains how to avoid the most common problems encountered with finish plumbing.

Sillcocks

If you asked the average person to give you a list of the plumbing fixtures installed in their home, they would never think to include sillcocks. While you might not think of a sillcock as a fixture, the plumbing code does. Sillcocks are one of the few fixtures that can be installed during the rough-in stage. In many cases, sillcocks are not installed until the end of the job. The factor that determines when they can be installed is accessibility. If you will not have access to the sillcocks when the house is finished, you must install them during the rough-in.

Sillcocks are one of the easiest plumbing fixtures to install. The first installation step is choosing a suitable location. Once you know the location, drill a hole for the sillcock. If the sillcock is being installed on the side of a frame house, make sure the siding has been installed on the home before you put in the sillcock. Drill the hole through the siding and wall sheathing. If the sillcock will be mounted to a masonry wall, use a hammer drill or star drill to make the hole in the masonry wall.

Sillcocks have a mounting flange that secures the sillcock to the wall with two fasteners (Fig. 10-1). Use screws to affix the sillcock to wood. Use plastic or lead anchors and screws to attach the sillcock to a masonry wall. If you are installing a frost-free sillcock, make sure the end of the sillcock extends past the exterior wall. Ideally, it should extend into an area that will be heated. In unheated areas, insulate the pipe to comply with the plumbing code.

Fig. 10-1. Sillcock.

Once the back end of the sillcock is in the desired location, connect the water supply to the sillcock. You should install a stop-and-waste valve in the pipe feeding the sillcock. Install the valve so that the arrow on the side of the valve points in the direction of the sillcock. Most sillcocks allow you to sweat a piece of copper into the back end of them. If you prefer, you can make the connection with a female adapter. Sillcocks generally have external threads to accommodate a female adapter in the connection. In cold climates, it is advisable to use frost-free sillcocks (Fig. 10-2).

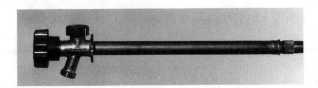

Fig. 10-2. Frost-free sillcock.

Bathtubs and showers

Bathtubs are installed during the rough-in stage of the job. How the tub is installed will depend on the type of tub being used. The two installations are one-piece shower bath combinations and bathtubs without a shower surround.

One-piece tub and shower combinations

One-piece tub and shower units are equipped with a nailing flange and are usually made from fiberglass or an acrylic material. The combination unit is the most common of all bathing units installed in modern plumbing systems. Installing this unit is not difficult.

Slide the unit into the opening that has been built to receive the bathing unit. If you are using a standard 5-foot unit, the rough opening should allow at least an additional $1/4$ of an inch for the installation. A full $1/2$ inch of extra space makes the installation much easier. One person can install these units alone, but the process is easier when two people work together on the installation.

Once the unit is in the cavity, level it. Use a 4-foot level when you set the tub/shower combination. Lay the level on the flood-level rim of the tub and check the bubble. If the bubble shows the tub to be level from front to back, half the battle is over. If the level shows the tub to be out of level, you will have to make a judgement call. If the tub is pitching slightly forward, you may proceed with the installation. If the tub is pitching backwards, correct the problem. If you ignore a backward pitch, the tub will not drain effectively.

If the tub is way out of level, contact a carpenter for a level floor on which to install the unit. If the tub is only slightly out of plumb, you can correct the problem with sand. Place sand under the tub at the back of the bathing unit. The sand will raise the rear of the tub and bring it to a level position. Most sub-floors are relatively level and it's unlikely you'll have much problem with the front-to-back leveling of the tub.

Once you are level on the flood-level rim, put the level on the vertical edge of the unit. You will most likely have to make minor adjustments for this part of the installation. Move the base of the unit in or out to reach a level point. If the tub is difficult to move, be careful. Fiberglass and acrylic units can break easily. Don't force the unit; if the opening is properly sized, the resistance is due to a crooked approach. Make sure the unit is straight and push or pull as necessary. After a few attempts, the unit should be level on the vertical rise.

Once the unit is level on the vertical rise, double check the horizontal leveling point on the flood-level rim. If both portions of the unit are level, secure it to the stud walls. To secure the unit, place nails or screws through the nailing flange and into the studs (Fig. 10-3). Most plumbers use roofing nails for this part of the job. Place the nail or screw in the center of the nailing flange and insert it into the stud. The fasteners should be placed liberally around the nailing flange.

Once the nails or screws are in, the bathing unit is set. All that is left to connect is the drain. For tub/shower combos, you will use a tub waste and overflow as a drain. Personally, I like the type that is made of schedule 40 plastic and glues together. Some plumbers prefer brass drains with slip nuts. If you use a tub waste with slip nuts, you must provide a panel in the wall that will allow access to the piping. With glue-together joints, the access panel is not required. For the few of you that will be using drum traps, you must leave access to the clean-out on the trap.

Tub wastes go together easily, but it helps to have a second set of hands (Fig. 10-4). The first step is to install the tub shoe, also known as a drain ell (Fig. 10-5). The tub shoe has female threads and sits under the drain in the bottom of the tub. The shoe has a flange that presses against the bottom of the tub. There will be a rubber gasket in the tub-waste kit that fits between the flange and the bottom of the tub. The drain for the tub is usually chrome, and has male threads on the end of it.

Apply plumber's putty around the flange of the chrome drain (Fig. 10-6). Take the putty in your hand and roll it into a line. Wrap the line around the drain's flange. Apply pipe dope or teflon tape to the threads of the drain. If you

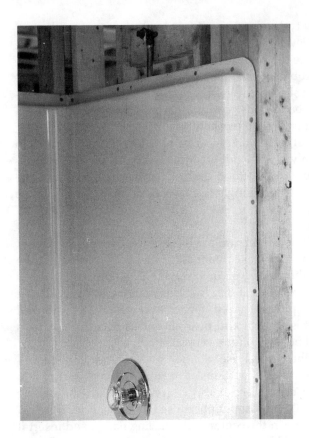

Fig. 10-3. Nailing flange of a one-piece tub/shower combination.

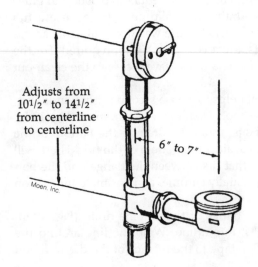

Adjusts from
10$^{1}/_{2}$″ to 14$^{1}/_{2}$″
from centerline
to centerline

6″ to 7″

Moen. Inc.

Fig. 10-4. A common tub waste.

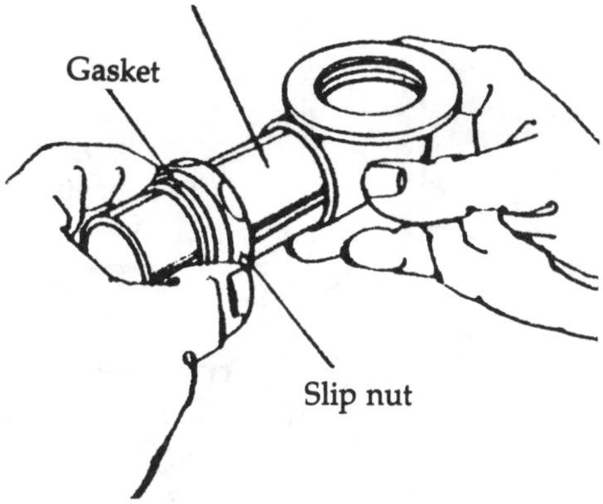

Drain tube

Gasket

Fig. 10-5. Installing the drain tube
of a tub waste. Moen. Inc.

Slip nut

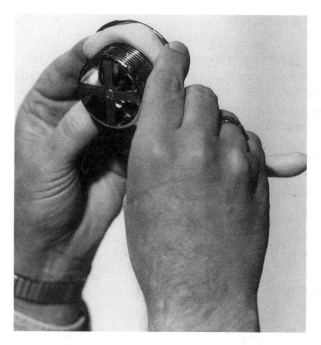

Fig. 10-6. Applying plumber's
putty to a tub drain.

use teflon tape, apply it so it will be tightened as the drain is turned clockwise. Hold the tub shoe under the drain with the gasket in place. Put the chrome drain through the drain hole and screw it clockwise into the tub shoe. When the connection is snug, point the end of the shoe's pipe towards the head of the tub.

Next, place the tee included with the tub-waste kit on the end of the shoe's pipe (Fig. 10-7). From this tee you will have a pipe that rises vertically to the overflow. The pipe should rise vertically in a nearly level manner. If the vertical pipe is cocked, cut the shoe's pipe to allow the tee to fit closer to the tub. This should allow the overflow pipe to rise upward without being cocked. In very rare circumstances, you might have to make the shoe's pipe longer to eliminate the

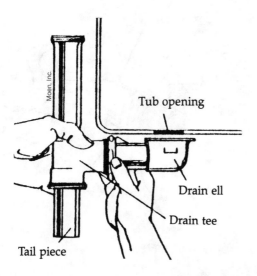

Fig. 10-7. Installing a tub shoe.

Tub opening

Drain ell

Drain tee

Tail piece

cocked overflow pipe. In all my years as a plumber, I cannot remember a situation when this was the case. Normally, the shoe's pipe will need to be shortened, if it needs to be adjusted at all.

Once you have the tee set, you are ready to install the overflow fittings and pipe (Figs. 10-8 and 10-9). The overflow portion of the tub drainage comes in two styles. The overflow fitting may be attached to the overflow pipe, or it may be a separate fitting. In schedule 40 plastic tub wastes, the overflow fitting will not usually have the overflow pipe attached. Brass tub wastes will have the overflow pipe attached to the fitting. Hold the overflow fitting in place at the overflow hole in the tub. Take a measurement that will allow you to cut the overflow pipe to the correct size.

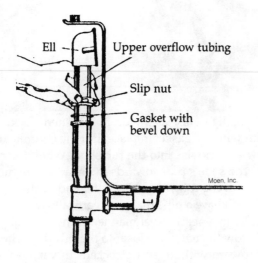

Ell

Upper overflow tubing

Slip nut

Gasket with bevel down

Fig. 10-8. Installing the overflow tubing of a tub waste.

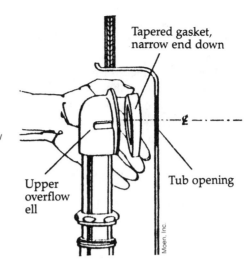

Fig. 10-9. Installing the overflow gasket of a tub waste.

Tapered gasket, narrow end down

Tub opening

Upper overflow ell

Moen, Inc.

Cut the overflow pipe and install it in the tee. If you are working with separate pieces, install the overflow fitting on the pipe and point it into the overflow hole of the tub. There will be a thick sponge gasket that goes between the flange on the overflow fitting and the back of the tub. Install this gasket on the overflow fitting.

The next connection will be between the tee and the tub's trap. With a brass tub waste, you have two options. You may screw a $1^1/2$-inch female adapter onto the external threads of the tee, or you can screw the supplied tailpiece into the internal threads. Apply pipe dope or teflon tape to the threads of the connection you make. When you are using plastic drainage pipe, I recommend using a female adapter for the connection. If you use a tailpiece connection, you will have to have the tailpiece connected to the trap with a trap adapter and a slip nut. This leaves the possibility of a leak in a concealed location.

When you use the female adapter, glue the joint between the trap and the tee. Either method is acceptable, but you will need an access panel if you opt for the slip-nut connection. When all of the other connections are made, tighten the tub drain into the tub shoe. Since most tub drains are equipped with cross-bar strainers, use two screwdrivers for this task. Cross the screwdrivers between the strainer and turn the drain clockwise until the putty under the rim spreads out evenly. Once the tub waste is connected to the trap, you are ready to install the trim pieces on the interior of the tub.

The trim-out parts will vary with different tub wastes. Most glue-together wastes will use a twist-and-turn stopper, or a push-up stopper. With these stoppers, the overflow plate is generally a solid chrome plate that is attached to the overflow flange with one or two screws. There will be an opening between the surface of the tub and the face plate to allow excess water to escape down the overflow pipe. When you install the face plate, the screws should be tightened until they compress the sponge gasket on the back side of the tub.

Installing the drain stopper for these units is a matter of screwing it into the tub shoe. There is no need for a pipe sealant on the threads of the stopper. All you do is screw it into the shoe.

With the push-type stopper, push it once to depress it and hold water in the tub. Depress it a second time to release the water in the tub. With the lift-and-turn style stopper, you lift the stopper, turn it, and allow it to seat in the drain. This procedure holds water in the tub. Lifting and turning the stopper again allows the tub's water to drain. I think these two types of tub wastes are the best available.

If you choose to use a trip-lever waste, the installation will be a little different. For these units, the opening and closing of the tub's drain is controlled with a lever. The lever is mounted in the face plate of the overflow fitting. Two screws hold this plate in place. When you unpack the tub waste, you will see the chrome face plate with a long rod extending from it. These rods are adjustable for tubs with different heights. Most tub wastes will be preset for standard tubs.

If the tub is extra deep or unusually shallow, you may have to adjust the rod attached to the overflow's face plate. This is done by loosening the retainer nut and turning the rod. Turn the rod counterclockwise to lengthen it and clockwise to shorten it. To install the face plate on a trip-lever waste, feed the rod into the overflow pipe. When the rod is all the way into the pipe, screw the face plate onto the overflow flange.

Depending on the style of trip-lever waste you use, you may have either of two types of drains. The first type will be a cross-bar strainer with a mesh strainer that attaches to the tub shoe. The other type will have a stopper plug that has a crooked piece of flat metal attached to it. With the mesh-strainer type, all you have to do is screw the mesh strainer to the tub shoe. For the other type, feed the flat metal section into the tub drain. When it is all the way into the shoe's drain, the stopper will be sitting in the proper position. If either of these types of tub wastes fail to hold water in the tub, adjust the length of the rod attached to the lever.

Installing the tub/shower faucet is your next step (Fig. 10-10). Chapter 7 gave you standard rough-in dimensions for the tub/shower faucet. The only part of the installation not covered in Chapter 7 was the process of cutting the holes in the tub/shower walls. The holes for the faucets and tub spout are best cut with a hole saw. Various drill bits will do the job, but a hole saw will give you the neatest hole.

If you are installing a single-handle faucet, the hole needed for the installation will be larger than any standard hole saw. For this situation, mark the size of the hole on the wall with a pencil. Drill a hole in the center of the marking. From the pilot hole, cut the larger hole with a reciprocating saw or a hacksaw blade. Be sure not to cut the hole larger than the escutcheon that will cover it.

Fig. 10-10. A typical mounting of a tub/shower faucet in a one-piece tub/shower combination.

Bathtubs without a shower surround

The waste and water connections on these tubs will go in the same as explained previously. The only big difference is how the tub is installed. This type of tub does not have the same type of nailing flange to secure to the stud walls. Standard bathtubs require the installation of a support system. The support is usually a piece of wood, like a 2-×-4 stud.

The wooden support should be about 58 inches long. It will be nailed to the studs so that the inside edge of the tub sits on the support. To determine the proper height for the support, you will have to measure the tub. With the tub sitting on the sub-floor, hold it up until it is level from the apron to the inside edge; this is the width of the tub, not the length. Measure from the bottom, inside edge of the tub to the sub-floor. Most tubs will measure between 13 and 14 inches. The height of your measurement will determine the top of the support.

If there are $13^1/_2$ inches between the floor and the edge of the tub, the support will be installed so that the top of it is $13^1/_2$ inches above the floor. This way, when the tub is resting on the support, it will be level from the back wall to the apron. When you nail the support in place, be sure it is level along its length. When the tub is set in place it should be level from front to back and side to side.

Once the tub is sitting on the support, check to be sure the tub is level. Cast-iron tubs are heavy enough to hold themselves in place. Steel tubs should be secured to the studs at the top of their drywall flange. Some steel tubs will be drilled with holes around the flange to allow you to nail or screw the tub to the wall. If there are no holes, drive roofing nails into the studs, just above the flange. The big head on the roofing nail will come down over the flange and secure the tub. Fiberglass tubs will be attached to the studs in the same manner as steel tubs.

One-piece shower stalls

The installation of a fiberglass or acrylic shower stall will go much the same way as a tub/shower unit. Set the unit in place and level it horizontally and vertically. Cut holes for the faucets as you would for tub/shower combinations. The faucet height for a shower is generally 4 feet above the sub-floor. The shower head height is $6^{1}/_{2}$ feet above the sub-floor. The big difference between installing a shower stall and a tub/shower unit is the drainage piping.

Shower stalls do not use tub wastes; they use a shower drain. The shower drain is easy to install, but the job goes better if you have some help. Put a ring of putty around the bottom of the shower drain's rim. Push the drain through the drain opening in the shower base. From below the shower, slide the fiber gasket onto the threaded portion of the drain. Next, install the large nut that came with the drain onto the threads. Turn the nut clockwise until it is tight. This is where it is advantageous to have help. When possible, have someone hold the drain in the shower, so the drain will not turn as you tighten the nut.

Once the drain is installed in the shower, you may connect it to the shower's trap. Most shower drains will be made from the same plastic pipe that you are using for the drain-waste-vent (DWV) system. The pipe size should be 2-inch. All you have to do to make the connection is install a piece of pipe between the trap and the drain.

Shower bases

There may be a time when you will use a shower base instead of a one-piece stall. If you plan to tile the shower walls, you will use only a base. Shower bases are a little different to install. Once you have the base sitting in the proper position and level, use roofing nails to hold it in place. Don't drive the nails into the base; drive the nails into the studs just above the base. The head on the roofing nail is large enough to come down over the base and hold it in place when the nail is driven into the stud wall.

The shower base drain might be different from the type used on a stall. Some bases come with the drain molded into the base. This type of drain has a metal collar extending down from the base. You may connect the drain pipe to the shower drain in two ways. The most common method employs the use of a rubber gasket. The gasket is put into the metal collar of the shower drain. Then the pipe is pressed up from below and into the rubber gasket. Another approach is to have the pipe protrude into the collar so that the gasket can be driven down onto the pipe.

In either case, the rubber gasket will need to be well lubricated to receive the pipe. You will often have to drive the gasket onto the pipe with a hammer. Installing these rubber gaskets is easier when you have help. It might be difficult to get the pipe into the gasket. With a second set of hands, the job will go smoother.

The second option for this type of drain is to seal the shower drain with molten lead. This is not a job for a homeowner or inexperienced plumber. The molten lead is extremely hot and may cause severe damage if it comes in contact with the human body. Due to the risk of working with hot lead, I will not elaborate on the methods used when working with lead. Without professional training and experience, you should not work with hot lead.

Toilets

Installing a toilet is not difficult. The job can be done by a professional in 15 minutes if the rough-in work was done correctly. When you prepare for the toilet installation, you should see a pipe for the water supply and a closet flange. These items were installed during the rough-in (Fig. 10-11). The first step is to install the closet bolts. These are the bolts that slide under the rim of the flange and protrude up through the holes in the base of the toilet. Closet bolts do not come with new toilets; you must purchase them separately. Place the wide, flat base of the bolts into the holes of the flange on each side of the drain. Slide the bolts into the groove until they are an equal distance from the back wall and in line with the center of the drain.

Fig. 10-11. Closet flange with closet bolts and wax ring.

Take a wax ring and place it on the flange. The wax ring is another item you must purchase separately. The ring should go directly over the drain, but it must not block the opening of the drain. Wax rings should be warm when they are installed. If your job site is unheated and cold, warm the wax ring with your vehicle's heater before you install it. The next step is to place the toilet bowl over the bolts and onto the wax ring. Press down firmly on the bowl to compress the wax ring.

When the bowl is sitting with its base on the floor, you must align the bowl with the back wall. Measure from the back wall to the holes in the bowl that accept the bolts from the toilet seat. When these holes are an equal distance from the back wall, install the nuts on the closet bolts. With some brands of toilets, you will install a plastic disc on the bolts before you put on the nuts. These disks allow you to snap cover caps in place later.

Turn the nuts clockwise with your hand until they are snug. Use an adjustable wrench to tighten the nuts, but don't get carried away. If you tighten these bolts too much, the toilet will crack. To avoid stress on any one of the nuts, alternate from one nut to the other as you tighten them. When the toilet bowl will not easily twist to either side, the nuts are tight.

Normally, once the nuts are tight the bolts will need to be cut. Cut the bolts with a hacksaw. When the bolts are short enough, install the cover caps to hide the bolts and nuts. If you had to install plastic disks on the bolts, the cover caps should snap into place on the disks. If you did not have any disks, put putty in the cover caps and press them onto the nuts. The putty will hold the caps in place.

If you are installing a one-piece toilet, you are ready to connect the water supply. If you are installing a standard, two-piece toilet, your next step will be the installation of the tank. All the parts needed to install the tank should have been provided with the tank. Take the large sponge or rubber gasket and place it over the threaded portion of the flush valve on the bottom of the tank. The bolts that hold the tank onto the bowl are called tank-to-bowl bolts. Normally there are two tank-to-bowl bolts, but some toilets require three.

The tank-to-bowl bolts will be packaged with the bolts, nuts, metal washers, and rubber washers (Fig. 10-12). First, slide the rubber washers onto the bolts until they rest against the head of the bolt. Sit the toilet tank onto the bowl with the flush-valve gasket sitting on the flush hole of the tank. It helps if you have someone hold the tank while you connect it to the bowl. If you are working alone, be careful not to let the tank fall. The china will break if it hits a hard surface. When the tank is in place, insert the tank-to-bowl bolts into the holes at the base of the tank.

Fig. 10-12. Tank-to-bowl bolts, nuts, and washers.

The bolts will go through the holes in the tank and bowl. Place the metal washers on the bolts and follow them with the nuts. Gradually tighten the nuts by hand. Alternate between the nuts when tightening them to avoid stress on any single area. Don't over-tighten these nuts; the tank will crack from the pressure. When the nuts are reasonably tight, move on to the next step of the installation. After the installation is complete, the nuts may be tightened if they leak.

All that is left is the seat and the water connection. To install the toilet seat, put it in place with the seat bolts going through the holes behind the seat. Install the nuts on the bolts, and the seat will be installed. Seats do not come with new toilets; they must be purchased separately.

For the water connection, make sure the water is turned off to the supply pipe for the toilet. At this stage of plumbing, the main cut-off valve is usually the only way to cut off the water to the supply pipe. Once the water is off, cut the supply pipe to install a cut-off on it. If the supply pipe is coming out of the wall, use an angle stop (Fig. 10-13). If the supply is coming out of the floor, use a straight stop (Fig. 10-14). Attach the cut-off using the same methods described for installing the pipe (Fig. 10-15). Before you install the stop, place an escutcheon on the supply pipe (Fig. 10-16). When copper pipe is used for the supply, it is not uncommon to use a compression stop. Using compression stops

Fig. 10-13. Compression angle stop and closet supply.

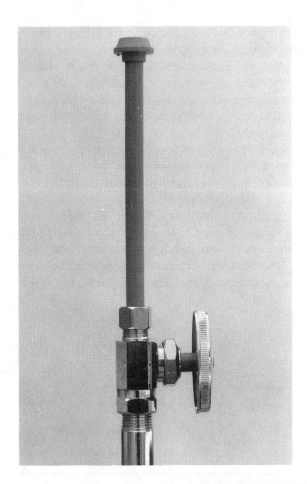

Fig. 10-14. Compression straight stop and closet supply.

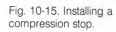

Fig. 10-15. Installing a compression stop.

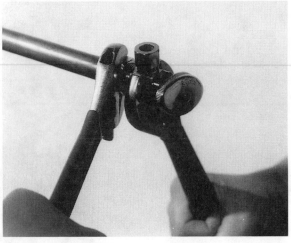

Fig. 10-16. An escutcheon around a supply pipe.

eliminates the need to solder the stops onto the supply pipes. This helps keep an open flame away from finished walls and cabinets.

To install a compression stop, slide the large compression nut onto the pipe. Next, slide the large compression sleeve onto the pipe. Then, put the stop onto the pipe. Push the sleeve up against the stop and screw the compression nut onto the stop. Hold the stop with an adjustable wrench as you tighten the nut with another wrench. There is no need to apply pipe dope to the threads of a compression fitting. The joint is made when the compression sleeve is compressed and held in place by the compression nut. When I use copper pipe, I use compression stops.

After the stop is affixed, install the closet supply (Fig. 10-17). This supply tube will have a flat head with a washer on it. Most modern, metallic closet supplies come with a plastic washer built onto the supply. Polybutylene supply tubes will be molded so that a washer is not needed. The stop and closet supply do not come with the new toilet. They must be purchased separately. You will have to cut the supply tube to the desired length. Metallic supply tubes are best cut with roller cutters, polybutylene supplies may be cut with a hacksaw.

Hold the head of the closet supply to the inlet of the ballcock. Bend the tube as needed to make it line up with the stop. When you bend metallic supplies, be careful not to crimp the tubing. You may want to buy a spring bender to bend the tubing. Professional plumbers can bend the tubing with their hands, but inexperienced people often crimp the tubing. Spring benders reduce the risk of damaging the tubing during the bending process. When the supply is in the right shape, cut it to the proper length.

Slide the ballcock nut onto the supply tube. The threads should point up towards the threads of the ballcock inlet. Slide the small compression nut and sleeve onto the supply tube with the threads facing the stop. You do not have to apply pipe dope to any of these threads. Hold the closet supply against the ballcock inlet and tighten the ballcock nut. When the nut is hand-tight, put the supply tube into the stop and tighten the compression nut. When everything is properly aligned, tighten both connections with an adjustable wrench. At this point, you have completed the toilet installation.

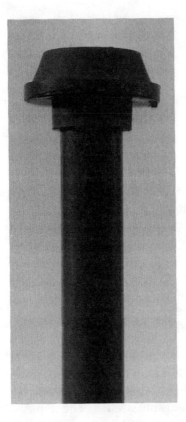

Fig. 10-17. A closet supply.

Lavatories

While there are many types of lavatories, the plumbing for the various types is essentially the same. Most professional plumbers trim-out the lavatory bowl before installing it. This saves time and the working conditions are better. There are two primary types of lavatory faucets. The first type has both the handles and the spout built into a single unit. The second type has the three pieces as separate units; each is installed independently, and then connected together. One-piece faucets are the most common.

One-piece lavatory faucets

These faucets are the easiest type to install (Fig. 10-18). Take the lavatory bowl and sit it on a counter or the floor. The faucet will have a base with two threaded inlets that extend below the base. If the faucet has a gasket to fit the base of the faucet, place it over the threaded inlets and against the faucet base. If you don't have a gasket, make one from plumber's putty. Roll the putty into a

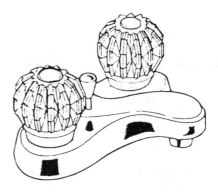

Fig. 10-18. A single-body, two-handle lavatory faucet. Universal-Rundal Corp.

long string and wrap it around the edges of the faucet base. Place the faucet on the lavatory with the inlets going through the holes in the bowl. There will be three holes in the bowl, the center hole will be used later.

Press the faucets onto the bowl to compress the putty, if the faucet did not come with a gasket. Take the ridged washers and place them on the threaded inlets from below the lavatory. Follow the washers with the mounting nuts. Tighten these nuts until the faucet is held firmly in place. If you plan to install faucets that do not have a common body, the procedure is different.

Three-piece faucets

Three-piece faucets require you to install each handle and the spout separately (Fig. 10-19). After they are all mounted, connect them with small tubing. There are many variations of 3-piece faucets. Most of them require you to place a nut and metal washer on the body of each unit. Then you push the units up, one at a time, through the holes in the lavatory and screw a mounting flange onto the body of the faucet. The flange is followed by a collar and the handle for the faucet. The spout will usually be held in place by a mounting nut below the bowl.

Fig. 10-19. A three-handle lavatory faucet. Universal-Rundal Corp.

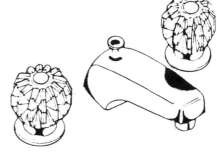

The spout will have an inlet on each side for hot and cold water. Connect a tube from each faucet body to the spout. This tubing is very sensitive and crimps easily. This type of faucet is more time consuming to install than a single-body faucet. In most cases, the tubing connection is made with compression fittings. Some models have the tubing soldered to the spout. Since there are so many possibilities, refer to the instructions for exact installation details.

The pop-up assembly

Now you are ready to install the drain. The drain for a lavatory is frequently called a pop-up (Fig. 10-20). When you look at the pop-up, there will be a number of pieces. The first thing you should do is unscrew the finished trim piece of the pop-up. Place a ring of putty around the bottom of this piece. The body of the pop-up will have a few inches of threads running down its length. There should be a large nut on these threads. Spin it down to the end of the threads near the middle of the pop-up body. Place the metal washer over these threads. Then, slide the beveled, rubber washer onto the threads. The bevel should be

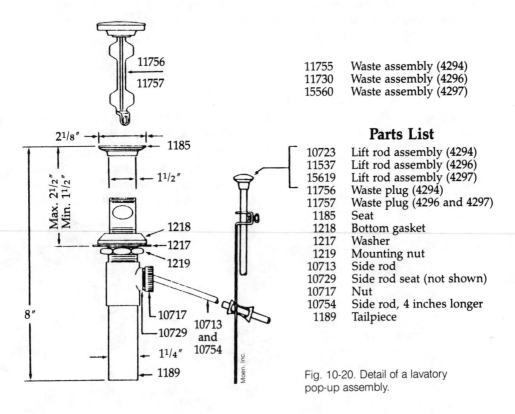

11755	Waste assembly (4294)
11730	Waste assembly (4296)
15560	Waste assembly (4297)

Parts List

10723	Lift rod assembly (4294)
11537	Lift rod assembly (4296)
15619	Lift rod assembly (4297)
11756	Waste plug (4294)
11757	Waste plug (4296 and 4297)
1185	Seat
1218	Bottom gasket
1217	Washer
1219	Mounting nut
10713	Side rod
10729	Side rod seat (not shown)
10717	Nut
10754	Side rod, 4 inches longer
1189	Tailpiece

Fig. 10-20. Detail of a lavatory pop-up assembly.

pointing up towards the bottom of the lavatory bowl. Apply pipe dope to the first few threads on the end of the body. Put the threads through the bowl's drain hole from underneath. Screw the finished part of the drain onto the threads. Don't use a wrench. A hand-tight installation is all that is required.

Pull the body down to seat the trim piece and putty against the bowl. Slide the beveled washer up to the drain hole. Tighten the large nut until the metal washer comes into firm contact with the beveled washer. Check the alignment of the finished drain and complete the tightening process. Tighten the nut until the putty spreads out and the beveled washer compresses.

There will be a tailpiece with the pop-up assembly. The tailpiece will be about 4 inches long and will have thin threads on one end. Apply pipe dope to the threads and screw the tailpiece into the bottom of the pop-up body. A hand-tight installation is all that is required.

There should be a rod extending from the back of the pop-up body. If this rod has not been installed, install it now. Place the pop-up plug in the drain from inside the lavatory bowl. The pop-up rod will have a nylon ball on the end of it. If there is a knurled nut on the threads where the rod will go, remove it. Slide the nut down the rod in a way that will allow it to screw onto the drain assembly and hold the rod in place. Insert the rod into the assembly and tighten the knurled nut. You may have to use a pair of pliers to make a tight connection, but the fitting usually will not leak if it is hand-tight.

The last step in the pop-up installation is the lift rod. Take the round rod and screw the head onto it. Slide the rod through the hole in the faucet between the handles. Take the perforated metal strip and put the rod into the hole in the top of the strip where the set-screw is. Take the thin metal clip and slide one end of it over the pop-up rod. The clip should remain near the end of this rod. Now, place the perforated metal strip so that one of its holes slides over the pop-up rod. Bend and place the remaining end of the clip onto the pop-up rod to hold the metal strip in place. Tighten the set-screw to secure the lift rod to the metal strip. As you pull up on the lift rod, the pop-up plug should go down to seal the drain. When you push the rod back down, the pop-up plug should come up and allow the drain to open. You may have to try various settings to obtain the proper adjustment between these connections.

Supply tubes

Before you set the lavatory, attach the lavatory supplies to the faucets (Fig. 10-21). Place the provided nuts on the supply tubes. Put the head of each supply tube into an inlet of the faucet body. Screw the supply-tube nut to the threaded portion of the supply inlet. If you do not install the supply tubes now, you will need a basin wrench to tighten the nuts after the lavatory is set.

Fig. 10-21. Lavatory supply tube.

Wall-hung lavatories

With all of this done, you are ready to install the lavatory bowl. Each type of lavatory will install a little differently. For a wall-hung lavatory, the first step is to install the wall bracket (Fig. 10-22). The bracket is screwed to the wall. A piece of wood backing should have been installed in the wall during the rough-in. Refer to your rough-in book for the proper height of the bracket. When the bracket is level and secure, place the lavatory on the bracket. Press down on the back rim of the bowl to seat it onto the bracket. Some models will have flanges and holes

Fig. 10-22. Wall-hung lavatory.

below the lavatory to allow the installation of extra screws. These screws are a safety precaution to prevent the lavatory from being knocked off the bracket. That is all there is to installing a wall-hung lavatory.

Rimmed lavatories

These units are not used much anymore (Fig. 10-23). If you are using a rimmed lavatory, the first step is to cut a hole in the countertop. The lavatory should have a template with it to show you how to cut the hole. Once the hole is cut, place the metal ring in the hole. To mount the bowl, push it up from below the counter until it touches the ring. Sink clips will be provided to mount the sink. These clips attach to the ring and hold pressure on the bowl as they are tightened. There will be installation instructions with the clips.

Fig. 10-23. Rimmed lavatory.

American Standard, Inc.

Self-rimming lavatories

For these units, use the supplied template to cut a hole in the countertop (Fig. 10-24). Apply a caulking sealant around the edge of the hole and sit the bowl into the hole. The weight of the bowl and the plumbing connections are all that hold this type of lavatory in place.

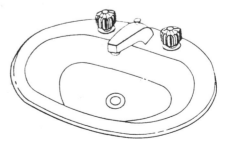

Fig. 10-24. Self-rimming lavatory.
Universal-Rundal Corp.

Molded lavatories

When using a top with the bowl molded into it, all you have to do is place the top on the cabinet (Fig. 10-25). The lavatory bowl is an integral part of the countertop and requires no additional installation.

Fig. 10-25. Molded lavatory in vanity top.

Pedestal lavatories

Pedestal lavatories are the most difficult of all for the average person to install (Fig. 10-26). The bowl is mounted to a wall bracket, just like a wall-hung lavatory.

Fig. 10-26. Pedestal lavatory.
Universal-Rundal Corp.

The pedestal is then placed under the bowl to hide piping and to help support the bowl. The complicated part of a pedestal sink is making the trap and supply connections. There is very little room to work and a minor miscalculation will ruin the effect of the pedestal. Some pedestals have a hole in the base to allow you to screw them to the floor. Others depend on the weight of the bowl to hold the pedestal in place. With any of these specialty fixtures, refer to the installation instructions and rough-in book.

The drain and trap

The trap will be attached to the trap arm and the lavatory tailpiece (Fig. 10-27). How these connections are made will depend on the type of trap you use. I will assume you are using a schedule 40 plastic drainage system. The trap may be glued to the trap arm if you use a schedule 40 plastic trap. If you prefer to use a chrome trap, you will need to make the trap-arm connection with a trap adapter. The trap adapter looks like a male adapter with a slip nut and washer on it. The trap adapter is glued to the trap arm and allows the connection between the adapter and the trap to be made with a slip nut and washer.

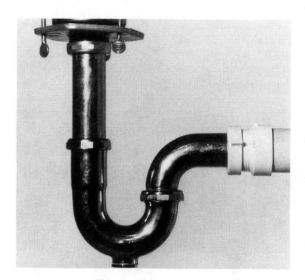

Fig. 10-27. Typical P-trap installation.

The connection between the trap and the tailpiece will always be made with a slip nut and washer. A lavatory tailpiece has a diameter of $1^1/4$ inches. If the trap is designed for a $1^1/2$-inch tailpiece, you may still use it. To modify the trap, all you need is a reducing washer for the slip nut. These reducing washers convert the $1^1/2$-inch trap opening to a $1^1/4$-inch opening.

To make the slip-nut connection, slide the nut up onto the tailpiece. Slide the washer onto the tailpiece and place the tailpiece in the trap. Slide the washer and nut down to the trap and tighten the nut. When the washer is compressed, the joint is made. With the use of trap adapters, you may use any standard trap to make the connection. If the trap is too low to connect to the tailpiece, you may use a tailpiece extension. These handy items come in various lengths and allow you to extend the length of the tailpiece. They connect to the tailpiece with a slip nut and washer.

Water supplies

Make sure the water to the water pipes is turned off. Follow the instructions given under the toilet section to install the cut-off stops. Use the same instructions given for the closet supply to install the lavatory supplies. The supply tubes will have a different type of head, but the installation methods are about the same. The lavatory supply will have a head tapered to fit into the inlets of the faucet body. There will be a nut that slides up the supply tube and screws onto the threaded portion of the inlets. These nuts hold the supply tubes in place. You do not need to apply pipe dope to the inlet connections for the supply tubes. Also, slip-nut connections do not require the use of pipe dope.

Kitchen sinks

As was the case with lavatories, kitchen sinks can be trimmed out before they are placed in the countertop. The faucets for kitchen sinks are usually of the single-body type (Figs. 10-28 and 10-29).

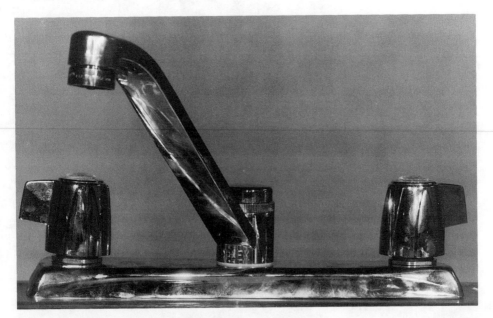

Fig. 10-28. One-piece, two-handle kitchen faucet.

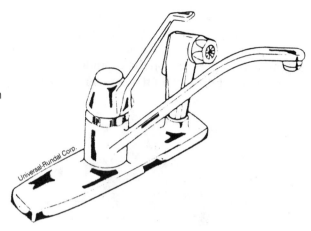

Fig. 10-29. Single-handle kitchen faucet with deck-mounted spray attachment.

The faucets

The faucets are installed following the same procedures for lavatory faucets. If the faucet has a spray-hose attachment, the hose will connect to a tapped opening in the center of the faucet's base. To install the sprayer, start with the housing for the unit.

The housing will be inserted in the hole next to the faucet base. Some faucets are designed for the hose to retract directly through the faucet base, but most have a separate housing for the hose. Once you have put the housing in the hole, screw the mounting nut onto the threads from below the sink. Feed the hose through the housing, with the brass threads going to the base of the faucet. Apply pipe dope to the threads and screw the hose connection into the female threads on the base of the faucet. If the spray head is not already attached, screw it onto the end of the hose above the sink.

Installing the drain assembly

When you are not mounting a garbage disposer to the sink, a basket strainer will be used as the drain assembly. There are two common types of basket strainers. The first uses a large nut to secure the strainer to the sink. The second uses a flange with threaded rods to secure the drain; this is the easiest type to install.

With either type, take the drain apart to install it. Apply putty under the rim of the finished piece and insert it in the drain hole. Press the drain down to spread the putty (Fig. 10-30). If you are using the type with a large nut, slip the fiber washer over the threads of the drain from below the sink. Follow the washer with the big nut. Tighten the nut until the putty is spread out and the drain is tight. If you have trouble with the drain turning as you attempt to tighten the nut, don't be surprised. If this happens, cross two screwdrivers through the bar grids in the drain. This will allow someone to hold the drain in place while you tighten the nut.

When you use a drain with a flange, you don't have to worry about the drain turning during the tightening. The flange is placed over the threaded part of the drain and the threaded rods are tightened. As you tighten the rods, the flange assembly puts pressure on the drain to force a good seal. At this point, you are ready to install the sink in the countertop.

Fig. 10-30. Basket strainer spreads putty as it is tightened.

Installing the sink

Use the template supplied with the sink to cut a hole in the countertop. Apply a caulking sealant to the top edge of the hole and set the sink in place. If you are using a cast-iron sink, the weight of the sink will hold it in place. If you are using a stainless steel sink, you will have to secure the sink with clips. The clips will come with the sink; refer to the installation directions.

In most instances, the clips will slide into a channel on the bottom of the sink. Then, the shaft that the metal clip is attached to will be turned clockwise. As you tighten the shaft, the clip bites into the bottom of the countertop. The pressure from the clip pressing against the countertop pulls the sink down tight.

Connecting the water supplies

You will use the same steps to connect the water supplies to the kitchen faucet that you employed for the lavatory faucet.

Connecting the drain

Most basket strainers will come with a flanged tailpiece. Unlike the lavatory tailpiece, these tailpieces do not screw into the drain. A tailpiece washer is placed on top of the tailpiece flange. A slip nut is slid up the tailpiece from the bottom. The tailpiece is positioned under the drain and the slip nut is tightened onto the drain's threads. Flanged tailpieces come in different lengths to adapt to the drainage system. Tailpiece extensions may be added, if needed, to lengthen the tailpiece.

On a single-bowl sink, the remainder of the connection between the trap arm and the drain is made in the same way as described for a lavatory. The only difference is the size of the drainage. Kitchen sinks use a 1½-inch drain. If you have a sink with more than one bowl, you may use a continuous waste to connect the multiple bowls to a single trap.

There are two styles of continuous wastes. The first is an end-outlet and the second is a center-outlet. The end-outlet brings the waste of one bowl to a tee under the other bowl (Fig. 10-31). A center-outlet brings the waste of both bowls to a tee in the center of the two bowls (Fig. 10-32).

Fig. 10-31. End-outlet continuous waste.

Fig. 10-32. Center-outlet continuous waste.

The placement of the trap arm will dictate the best type of continuous waste to use. Continuous wastes go together with slip nuts and washers. Once the bowls are connected by the continuous waste, the connection between the trap arm and the tailpiece of the waste will go together just like the waste for a lavatory.

Garbage disposers

If you are installing a garbage disposer, it will be attached to the drain of the sink bowl. The disposer eliminates the need for a basket strainer. The finished part of the disposer's drain will be installed like that of a basket strainer. From below, slide a flange over the drain. The flange will be held in place with a snap ring that fits on the drain collar. You will tighten threaded rods to secure the drain to the sink. When the drain is tight, the disposer will be mounted to the drain collar.

The mounting is usually done by holding the disposer on the collar and turning it clockwise. There will be a small, plastic elbow to install on the side of the disposer. You will see a hole in the side of the disposer with a metal ring attached to it by screws. Remove the metal ring and slide it over the elbow until it reaches the flange on the elbow. Place the supplied rubber gasket on the face of the elbow. Hold the elbow in place and attach the metal bracket to the side of the disposer with the supplied screws. The trap or continuous waste will attach to the elbow with slip nuts and washers.

Dishwashers

The dishwasher will have a $5/8$-inch drain hose attached to it. This drain should enter the cabinet under the kitchen sink and be attached to an airgap. An airgap is a device that mounts onto the sink or countertop and has a wye connection (Fig. 10-33). To install the airgap, remove the chrome cover by pulling on it. Remove the mounting nut from the threads. Push the airgap up through a hole in the sink or counter from below. Install the mounting nut and replace the chrome cover.

There will be two ridged connection points on the airgap. The first will accept the $5/8$-inch hose from the dishwasher. Before putting the hose on the airgap, slide a stainless steel clamp over the hose. Next, push the hose onto the ridged connection and tighten the clamp around the hose and insert connection. Attach a piece of $7/8$-inch hose to the other connection point using the same procedure.

When the $7/8$-inch hose leaves the airgap, it will attach to a garbage disposer or a wye tailpiece. The wye tailpiece attaches to the sink tailpiece like a tailpiece extension (Fig. 10-34). The hose will attach to the wye portion of the extended tailpiece and will be held in place with a stainless steel clamp. If you are connecting to a disposer, knock out the plug in the disposer first. There will be a thin

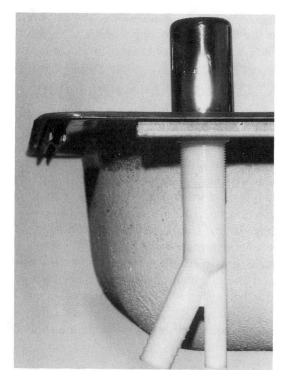

Fig. 10-33. Airgap mounted on kitchen sink.

Fig. 10-34. Wye tailpiece for connecting a dishwasher drain.

metal disk seated in the side of the disposer that must be removed. This is normally done with a hammer and a screwdriver. Once the plug is removed, attach the hose to the connection with a stainless steel clamp.

Connecting the water supply

There is a small box under the dishwasher where the water supply will be connected. Buy a dishwasher ell to make the connection. One end of this elbow has male threads and the other end is a compression fitting. Apply pipe dope to the threads and screw it into the dishwasher. You will need to tap into the hot water under the kitchen sink.

The hot water connection may be made with a tee fitting or a special type of cut-off stop. The special stops are designed to feed the sink and a dishwasher (Fig. 10-35). The connection point for the dishwasher tubing will be the correct size, without any type of adapter. If you install a tee fitting, the dishwasher must have its own cut-off valve. A stop-and-waste valve is the type normally used. If you use a 1/2-inch valve, you will need a reducing coupling. The tubing going to the dishwasher will have an inside diameter of 3/8 of an inch. The reducing coupling will be a 1/2-×-3/8-inch reducer. Connect the tubing to the reducer and run it to the dishwasher elbow. Make the connection at the compression fitting and you're done.

Fig. 10-35. Compression combination stop used in supplying water to a kitchen sink and dishwasher from the same valve.

Ice makers

When you have an ice maker to hook up, use a self-piercing saddle valve and 1/4-inch tubing. The tubing will connect at the back of the refrigerator with a 1/4-inch compression fitting. The tubing may run to the water supply at the kitchen sink or to another more accessible cold-water pipe. The saddle valve will clamp around the cold-water pipe and be held in place by two bolts (Fig. 10-36).

Fig. 10-36. Saddle valve.

Before bolting on the saddle, make sure the rubber gasket that came with the saddle is in place. It should be in the hollow of the saddle, where the piercing will take place on the pipe. Secure the saddle and connect the tubing to it. This connection will be made with a compression fitting. Turn the handle clockwise until you cannot turn it any farther. Then, turn the handle counterclockwise to open the valve and allow water to run through the tubing.

Water heaters

Water heater inlets may have male or female threads. The inlets will be marked to identify the hot and cold water connections. To connect a water heater, install the appropriate adapters on the inlets and connect the inlets to the water piping. You must have a gate valve on the incoming water pipe and there should be no valves on the outgoing water pipe. Most places require a vacuum breaker to be installed on the inlet pipe. A temperature and pressure relief (t&p) valve will be needed (Fig. 10-37).

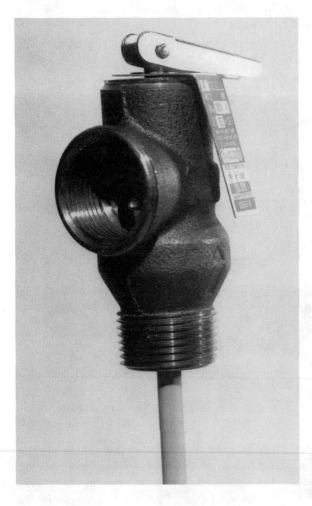

Fig. 10-37. Temperature and pressure relief valve for a water heater.

The t&p valve must be rated to safely operate with the water heater you are installing. The t&p valve will screw into the top or side of the water heater. A discharge pipe must run from the t&p valve to within 6 inches of the floor. Use pipe dope on all the threaded connections used to set a water heater.

The wind down

This concludes the basics of installing standard, household plumbing fixtures. In all cases, you should read the installation instructions provided with the fixtures and abide by them. This advice is based on general conditions. Your fixtures may require other installation procedures. If you are installing a specialty fixture, this information, combined with that of the manufacturer, should see you through the job.

11

Installing water pumps and conditioners

WHEN THE WATER comes from a well, water pumps are the heart of the plumbing system. When the pump fails, the potable water system is helpless. With the pump being critical to the successful operation of the water distribution system, it will pay big dividends to install it properly. Water conditioners are often needed to treat well water. They may be used to control acid in the water, minerals, or even bacteria. This chapter will cover the installation of potable water pumps and water conditioners.

Shallow-well jet pumps

Shallow-well pumps are used in wells where the lowest water level will not be more than 25 feet below the pump (Fig. 11-1). There are many factors that influence the type and size of pump the installation will require. The first consideration is the height to which the water will need to be pumped. If the pump will have to pump water higher than 25 feet, a shallow-well pump is not a viable choice. Shallow-well pumps are not intended to lift water more than 25 feet. This limitation is due to the way that a shallow-well pump works.

These pumps work on a suction principle. The pump sucks the water up the pipe and into the home. With a perfect vacuum at sea level, a shallow-well pump may be able to lift water to 30 feet. This maximum lift is not recommended and is rarely achieved. If you will have to lift the water higher than 25 feet, investigate other types of pumps. If you can use a shallow-well jet pump, this section will tell you how to install it. When you have questions about sizing the pump, talk to a pump dealer. He will be happy to size the pump for you.

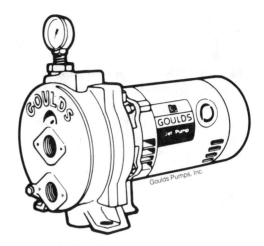

Fig. 11-1. Shallow well pump.

Goulds Pumps, Inc.

Exterior piping

When you install a suction pump, the single pipe from the well to the pump will usually have a diameter of $1^1/_4$ inches. A standard well pipe material is polyethylene, rated for 160 pounds per square inch (psi). You will want to be sure the suction pipe is not coiled or in any condition that may cause it to trap air. If the pipe holds an air pocket, priming the pump might be quite difficult.

In most cases, you should install a foot valve on the end of the pipe that is submerged in the well (Fig. 11-2). Screw a male insert adapter into the foot valve.

Fig. 11-2. Foot valve.
Goulds Pumps, Inc.

Place two stainless steel clamps over the well pipe and slide the insert fitting into the pipe. Tighten the clamps to secure the pipe to the insert fitting. When you lower the pipe and foot valve into the well, don't let the foot valve sit on the bottom of the well. If the suction pipe is too close to the bottom of the well, it may suck sand, sediment, or gravel into the foot valve. If this happens, the pipe cannot pull water from the well.

When the pipe reaches the upper portion of the well, it will usually take a 90-degree turn when it exits the well casing. This turn will be made with an insert-type elbow. Always use two clamps to hold the pipe to its fittings. When the pipe leaves the well, it should be buried underground. The pipe must be deep enough so that it will not freeze in the winter. This depth will vary from state to state. Your local plumbing inspector will be able to tell you how deep to bury the water supply pipe.

When you place the pipe in the trench, be careful not to lay it on sharp rocks or other objects that might wear a hole in the pipe. Backfill the trench with clean fill dirt. If you dump rocks and cluttered fill on the pipe, it might be crimped or cut. When you bring the pipe into the home, run it through a sleeve where it comes through or under the foundation. The sleeve should be two pipe sizes larger than the water supply pipe.

Interior piping

Once inside the home, you may wish to convert the pipe to copper, CPVC, or one of the other approved materials. If you convert the pipe, the conversion will typically be done with a male insert adapter. The water supply pipe should run directly to the pump. The foot valve acts as a strainer and check valve. When you have a foot valve in the well, there is no need for a check valve at the pump.

The incoming pipe will attach to the pump at the inlet opening with a male adapter (Fig. 11-3). At the outlet opening, install a short nipple and a tee fitting. At the top of the tee, install reducing bushings and a pressure gauge. From the center outlet of the tee, the pipe will run to another tee fitting. There should be a gate valve installed in this section of pipe, near the pump.

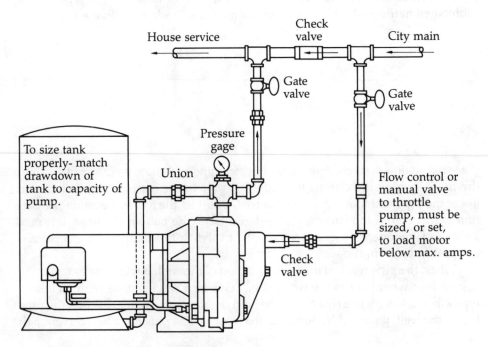

Fig. 11-3. Typical jet pump setup. Goulds Pumps, Inc.

At the next tee, the center outlet will be piped to a pressure tank. From the end outlet of the tee, the pipe will run to yet another tee fitting. At this tee, the center outlet will become the main cold-water pipe for the house. Another gate valve should be installed in the pipe feeding the water distribution system. On the end outlet of the tee, install a pressure-relief valve. All of these tee fittings should be in close proximity to the pressure tank.

Electrical connections

The pump will be equipped with a control box that requires electrical wiring. This job should be done by a licensed electrician. If you are an electrician, you know how to do the job. If you are not an electrician, you shouldn't attempt to wire the controls.

Priming the pump

There will be a removable plug in the top of the pump to allow you to prime it. Remove the plug and pour water into the priming hole. Continue this process until the water is standing in the pump and visible at the hole. Apply pipe dope to the plug and screw it back into the pump. When you turn the pump on, you should have water pressure. If you don't, you must continue the priming process until the pump is pumping water. This can be a time consuming process. Don't give up.

Setting the water pressure

Once the pump is pumping water, the pressure tank will fill. When the tank is filled, the pressure gauge should give a reading between 40 and 60 psi. The pump's controls will be preset at cut-in and cut-out intervals. These settings regulate when the pump cuts on and off. Typically, a pump will cut on when the tank pressure drops below 20 psi. The pump will cut off when the tank pressure reaches 40 psi.

If you prefer a higher water pressure, the pressure switch may be altered to deliver higher pressure. You might have the controls set to cut on at 40 psi and off at 60 psi. Adjusting these settings is done inside the pressure switch around electrical wires. There is the possible danger of electrocution when you make these adjustments. Unless you are experienced with such work, leave the adjustments to a licensed plumber or electrician.

The adjustments are made by turning a nut that sits on top of a spring in the control box (Fig. 11-4). If you attempt this, and I don't recommend that you do, cut off the power to the pressure switch before you open the control box. Once opened, you will see a coiled spring that is compressed with a retaining nut. You may alter the cut-in and cut-out intervals by moving this nut up and down the threaded shaft. The voltage from the wires in the pressure switch can deliver a fatal shock. Do not attempt this job unless you are experienced in such work.

Fig. 11-4. Cut-in, cut-out spring adjustments in pressure switch.

Deep-well jet pumps

When the water level is more than 25 feet below the pump, you will have to use either a deep-well jet pump or a submersible pump. In today's plumbing applications, submersible pumps are normally used in deep wells. However, deep-well jet pumps will get the job done and are covered in this section.

Deep-well jet pumps resemble shallow-well pumps (Fig. 11-5). They look about the same. They are installed above ground and are piped in a manner similar to shallow-well pumps. The noticeable difference is the number of pipes going into the well. A shallow-well pump has only one pipe. Deep-well jet pumps have two pipes. The operating principles of the two types of pumps differ. Shallow-well pumps suck water up from the well. Deep-well jet pumps push water down one pipe and suck water up the other. This is why there are two pipes on deep-well jet pumps.

The only major installation differences between a shallow-well pump and a deep-well pump are the number of pipes used in the installation and the pressure control valve. Deep-well pumps still use a foot valve. There is a jet-body fitting that is submerged in the well and attached to both the pipes and foot valve (Fig. 11-6). The pressure pipe connects to the jet assembly first. The foot valve

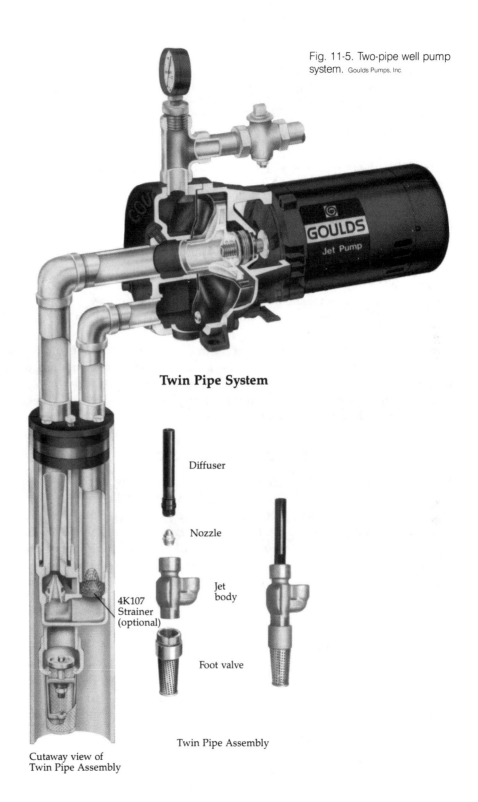

Fig. 11-5. Two-pipe well pump system. Goulds Pumps, Inc.

Twin Pipe System

Diffuser

Nozzle

Jet
body

4K107
Strainer
(optional)

Foot valve

Twin Pipe Assembly

Cutaway view of
Twin Pipe Assembly

Installing water pumps and conditioners 157

Fig. 11-6. Jet body.
Goulds Pumps, Inc.

hangs below the pressure pipe. There is a molded fitting on the jet body to which the suction line connects. With this jet body, both pipes are allowed to connect in a natural and efficient manner.

In order for the suction pipe to pull water up from the deep well, water is pushed through the jet assembly and down the pressure pipe. From the suction pipe, water is brought into the pump and distributed to the potable water system. When you look at the head of a deep-well jet pump, you will see two openings to which the pipes connect. The larger opening is for the suction pipe and the smaller opening is for the pressure pipe. The suction pipe will usually have a diameter of $1^{1}/_{4}$ inches. The pressure pipe will typically have a diameter of 1 inch.

You will need to install a pressure control valve on the piping running from the pump to the pressure tank. This valve assures a minimum operating pressure for the jet assembly. Shallow-well pumps do not require a pressure control valve. Once the pressure control valve is installed, the remainder of the piping is done in the manner used for a shallow-well pump.

Submersible pumps

Submersible pumps are very different from jet pumps (Fig. 11-7). Jet pumps are installed outside of the well. Submersible pumps are installed in the water in the well. Jet pumps use suction pipes. Submersible pumps have only one pipe and they push the water up the pipe. Jet pumps use a foot valve, submersible pumps don't. Submersible pumps are much more efficient than jet pumps. They are also easier to install. Under the same conditions, a $^{1}/_{2}$-horsepower submersible pump can produce nearly 300 gallons more water than a $^{1}/_{2}$-horsepower jet pump. With so many advantages, it is almost foolish to use a jet pump when you could use a submersible pump.

The installation of a submersible pump requires different techniques. Since submersible pumps are installed in the well, electrical wires must run down inside the well to the pump. Before you install the submersible pump, consult a licensed electrician about the pump's wiring needs. You will need a hole in the well casing in order to install a pitless adapter. The pitless adapter provides a watertight seal in the well casing through which the well pipe will run to the water service. When you purchase the pitless adapter, it should be packaged with instructions on what size hole you will need in the well casing.

You may cut a hole in the well casing with a cutting torch or a hole saw (Fig.

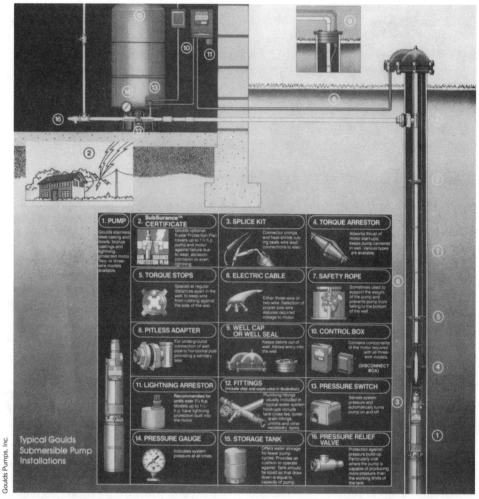

Typical Goulds
Submersible Pump
Installations

1. PUMP	2. SubSurance™ CERTIFICATE	3. SPLICE KIT	4. TORQUE ARRESTOR
Goulds stainless steel casing and bowls; bronze castings and lightning-protected motor. Two- or three-wire models available.	Goulds optional 5-year Protection Plan covers up to 1½ h.p. pump and motor against failure due to wear, abrasion, corrosion or even lightning.	Connector crimps and heat-shrink tubing seals wire lead connections to electric cable.	Absorbs thrust of motor start-ups; keeps pump centered in well. Various types are available.
5. TORQUE STOPS	6. ELECTRIC CABLE	7. SAFETY ROPE	
Spaced at regular distances apart in the well, to keep wire from rubbing against the side of the well.	Either three-wire or two-wire. Selection of proper size wire assures required voltage to motor.	Sometimes used to support the weight of the pump and prevents pump from falling to the bottom of the well.	
8. PITLESS ADAPTER	9. WELL CAP OR WELL SEAL	10. CONTROL BOX	
For underground connection of well pipe to horizontal pipe providing a sanitary seal.	Keeps debris out of well. Allows entry into the well.	Contains components of the motor required with all three-wire models. (DISCONNECT BOX)	
11. LIGHTNING ARRESTOR	12. FITTINGS (Include stop and waste valve in illustration)	13. PRESSURE SWITCH	
Recommended for units over 1½ h.p. Models up to 1½ h.p. have lightning protection built into the motor.	Plumbing fittings usually included in typical water system hook-ups include tank cross tee, boiler drain fittings, unions and other necessary items.	Senses system pressure and automatically turns pump on and off.	
14. PRESSURE GAUGE	15. STORAGE TANK	16. PRESSURE RELIEF VALVE	
Indicates system pressure at all times.	Offers water storage for fewer pump cycles. Provides air cushion to operate against. Tank should be sized so that draw down is equal to capacity of pump.	Protection against excessive pressure build-up. Particularly vital where the pump is capable of producing more pressure than the working limits of the tank.	

Fig. 11-7. Submersible pump.

11-8). The pitless adapter will attach to the well casing and seal the hole. On the inside of the well casing, attach a tee fitting onto the pitless adapter. This is where the well pipe will be attached. This tee fitting is designed to allow you to make all of the pump and pipe connections above ground. After all the connections are made, lower the pump and pipe into the well. The tee fitting will slide into a groove on the pitless adapter.

To make up the pump and pipe connections, you will need to know the depth of the well. The well driller should provide you with the well's depth and rate of recovery. Once you know the depth, cut a piece of plastic well pipe to the desired length. The pump should hang at least 10 feet above the bottom of the well and at least 10 feet below the lowest expected water level.

Installing water pumps and conditioners **159**

Fig. 11-8. Well casing and cap.

Apply pipe dope to a male-insert adapter and screw it into the pump. This fitting is normally made of brass. Slide a torque arrestor over the end of the pipe. Next, slide two stainless steel clamps over the pipe. Place the pipe over the insert adapter and tighten the two clamps. Compress the torque arrestor to a size slightly smaller than the well casing and secure it to the pipe. The torque arrestor absorbs thrust and vibrations from the pump and helps to keep the pump centered in the casing.

Slide torque stops down the pipe from the end opposite the pump. Space the torque stops at routine intervals along the pipe to prevent the pipe and wires from scraping against the casing during operation. Secure the electrical wiring to the well pipe at regular intervals to eliminate slack in the wires. Apply pipe dope to a brass, male-insert adapter and screw it into the bottom of the tee fitting for the pitless adapter. Slide two stainless steel clamps over the open end of the pipe and push the pipe onto the insert adapter and tighten the clamps.

Before you lower the pump into the well, it is a good idea to tie a safety rope to the pump. After the pump is installed, tie the rope to the top of the casing so that the pump is not lost if the pipe becomes disconnected from the pump. Next, screw a piece of pipe or an adapter into the top of the tee fitting for the pitless adapter. Most plumbers use a rigid piece of steel pipe for this purpose.

Once the pipe extends up from the top of the tee fitting, lower the whole pump assembly into the well casing. This job is easier if you have someone to help you. Be careful not to scrape the electrical wires on the well casing as the pump is lowered. If the insulation on the wires is damaged, the pump may not work. Hold the assembly by the pipe that extends from the top of the pitless tee and guide the pitless adapter into the groove of the adapter in the well casing. When the adapter is in the groove, push down to seat it into the mounting bracket. This concludes the installation inside the well.

Attach the water service pipe to the pitless adapter on the outside of the casing. You can do this with a male-insert adapter. Run the pipe to the house in the way described for jet pumps. Once inside the house, the water pipe should have a union installed in it. The next fittings should be a gate valve followed by a check valve. From the check valve, the pipe should run to a tank tee.

The tank tee is a device that screws into the pressure tank and allows the installation of all related parts. The switch box, pressure gauge, and boiler drain can all be installed on the tank tee. When the pipe comes to the tank tee, the water is dispersed to the pressure tank, drain valve, and water main. When the water main leaves the tank tee, install a tee to accommodate a pressure relief valve. After this tee, you may install a gate valve and continue the piping to the water distribution system. All that is left is to test the system. You do not have to prime a submersible pump.

Water conditioners

When it comes to water conditioners, there are numerous types of equipment from which to choose (Fig. 11-9). These water treatment devices are used to control and eliminate many elements found in water. The paragraphs below describe some of the most common factors found in drinking water, and filters used with water conditioning systems to correct the problem.

Hard water. Water conditioners may be used to remove hardness from the water supply. Hard water makes it more difficult to clean clothes, may leave a residue on the plumbing fixtures, and may lower water pressure. In addition, it may leave deposits in the pipes and water heater that might cause premature deterioration of the pipes and equipment. If you have ever had a dull film left on your dishes, you probably washed them in hard water.

High iron content. When the water has a high iron content, the plumbing fixtures will be discolored. Water with iron in it may have an unpleasant taste. If the fixtures contain a high slime content, you probably have a high iron content. Iron filters are used to remove large quantities of iron from the water.

Sulfur. If the water has a high sulfur content, you will know it by the smell of the water. The water will smell like rotten eggs. Water with sulfur in it can appear almost black. Sulfur water can have a corrosive effect on the pipes, fixtures, and equipment. Unpleasant tastes and odors are treated with carbon filters. Cloudy water may be cured with a sand filter.

Acidity. Water with a high acid content can damage the pipes, stain the fixtures, and affect your health. If you notice a blue-green stain on the fixtures, acidic water is present. The acid in the water will deteriorate copper pipes and plumbing equipment. If you have galvanized steel water pipes, the acidic water

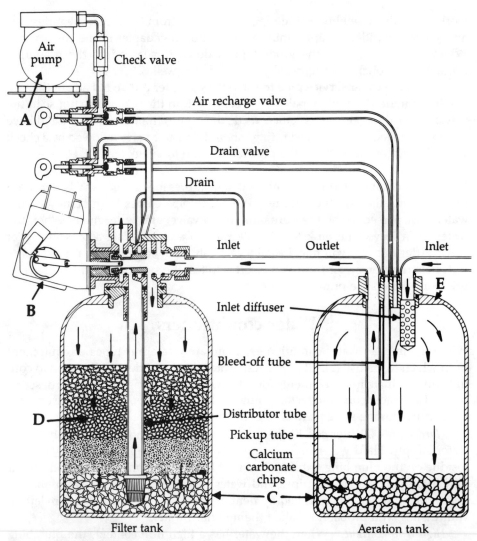

Air pump

Check valve

Air recharge valve

A

Drain valve

Drain

Inlet   Outlet   Inlet

B

Inlet diffuser

E

Bleed-off tube

Distributor tube

Pickup tube

Calcium
carbonate
chips

D

C

Filter tank   Aeration tank

Fig. 11-9. Water conditioner. Iron Curtain Filter System Parts Description; Source: Hellenbrand Water Conditioners, Inc., Waunakee, WI Iron Curtain Filter System Parts Description.

will cause them to rust. If your body is sensitive to acid, your health may suffer. Acid neutralizers may control the acidic content of the water. I once installed an acid neutralizer for a lady who was experiencing extreme discomfort from acidic water. After installing the acid neutralizer, her health returned to normal.

Diseases. Untreated water may contain a host of serious diseases. Typhoid fever is caused by impurities in the water. Other diseases include: paratyphoid, cholera, schistosomiasis, hepatitis, diarrhea, and others. Bacteria is often treated with chlorination and filtration. With the potential of such dangerous diseases, it is important to have all water from a private source tested for purity.

Water tests.   Water tests are usually required by the code enforcement office when a new plumbing system is installed. If the test is not required, you should perform one on your own initiative. Many places provide a free water analysis. Contact local plumbers, municipal offices, water treatment companies, well companies, etc., for information.

If you prefer to perform your own test, there are independent labs that will test the water for you. These professional laboratories will provide you with a container in which to collect a water sample. They will also give you instructions for how to take the sample. Check in your local phone directory for companies that test water.

Have the quality of the water evaluated. When you receive the report, determine the type of treatment and equipment needed. Water treatment equipment will be sold with installation instructions. If you feel confident you can handle the installation, closely follow the instructions. Some water conditioners are very sensitive. The wrong settings can render the equipment useless. Review the installation instructions, before you decide if you are qualified to do the job.

The piping to the equipment will be done with normal plumbing materials. The instructions will show the proper pipe installation methods. With a simple system, you should be able to handle the installation. With more complex systems, it often pays to have an expert make the installation.

Closing comments

This chapter dealt with private water supplies. We have covered everything from pumps to conditioners. Your health may be directly related to the quality of the potable water. Don't take shortcuts when you are dealing with such an important issue. Make all of the connections with the proper materials and in the proper way. Have the water tested. If the test results indicate a need for treatment, don't delay. Install the equipment needed to treat the water. Trying to save a few dollars might cost you your health.

12

Final testing and code inspection

ONCE ALL THE plumbing is installed, you will have to call for a final inspection. These inspections are similar to the inspection performed at the rough-in stage. Without an approved final inspection, you will not be able to obtain a certificate of occupancy for a new home. If you don't call for a final inspection of a remodeled job, the code office will contact you and require you to comply with the inspection rules.

Before you call for this final inspection, check all the fixtures and connections to be sure they will pass inspection. If your job fails its first inspection, the plumbing inspector is apt to look a little harder during the second inspection. After a preliminary check of the plumbing, there is no reason why your job should fail the official inspection. Let's work our way through the new installation and see how you may be able to avoid a rejection slip from the plumbing inspector.

Toilets

When the plumbing for the toilet was inspected during the rough-in inspection, the clearance requirements for the fixture were checked. Now the inspector will be looking at how you installed the toilet. Try to move the bowl from side to side. If the bowl shifts, tighten the closet bolts. The bowl should be secured to the flange to prevent it from moving. Next, check the tank for movement. The toilet tank should be firmly attached to the bowl. If you can rock the tank with gentle pressure, tighten the tank-to-bowl bolts. Remember to be careful when you tighten bolts in china fixtures. Too much stress will crack the fragile china.

Attempt to move the seat from left to right. If the seat is loose, tighten the nuts holding the seat to the bowl. Flush the toilet and watch as the water leaves the bowl. The water should leave with force. If the water swirls and goes down slowly, the inspector may reject the job. In new installations, the wax ring sometimes partially blocks the toilet's drain. This occurs when the bowl is being set. If the toilet drains slowly, try using a plunger and closet auger to clear any obstructions in the trap. If the water still moves slowly, lift the toilet and inspect the wax seal. Make sure the toilet has a strong flush before you call for the inspection.

After you flush the toilet, look inside the tank and check the water level. If the flapper or flush ball is not sealing, water will leak through the flush valve and into the bowl. The inspector will check to see that this type of leak is not present. You may have to adjust the flapper's chain or the tank ball's lift rods if water is running into the bowl after the tank is filled. The refill tube from the ballcock should be attached so that its stream of water enters the overflow tube.

Look at the stop valve and closet supply. It is not uncommon for compression fittings to work loose and leak after the initial installation. A simple drip from a compression fitting is all it takes to fail the inspection. Wipe all the water connections with a paper towel. Leaks that evade your vision will show up on the paper towel. Flush the toilet again. Then, with a paper towel, wipe around the area where the tank joins the bowl. This test will expose small leaks around the tank-to-bowl bolts and flush valve.

If the tank-to-bowl bolts leak, some gentle tightening will usually solve the problem. If the leak persists, look to the washers under the head of the tank bolts. If a leak persists and the tank-to-bowl bolts are tight, check the sponge gasket between the flush valve and the bowl. To do this, you will have to remove the tank from the bowl.

The last thing to test for is a leak around the base of the toilet. If the wax ring is not doing its job, water will seep out around the base of the toilet. This type of leak generally is caused by a poor installation of the wax ring, or a closet flange that was set too far below the finished floor. If you have water coming out around the toilet's base, you will have to remove the water closet from the flange.

Look at the wax ring to see if you can find a reason for the leak. If the wax ring is not compressed, the flange is too low (Fig. 12-1). The flange should be set so that the mounting surface is at, or near, the finished floor level. When the flange sits well below the finished floor, you will have to use two wax rings (Fig. 12-2). In this situation, use the wax rings with plastic horns (Fig. 12-3). Set the first wax ring in place. Then set another ring on top of the first one. When you double the wax rings, you make it possible for the closet bowl to compress the wax and make a seal.

If the flange is set at the right height, the leak is probably due to a poor installation of the wax ring. Install a new wax ring and reset the toilet. Be careful to bring the bowl down evenly on the wax. Again, wax rings with plastic horns are always a good choice. As long as the wax is spreading out between the flange and the bowl, you should get a solid seal when you evenly set the bowl.

Fig. 12-1. Closet flange with wax ring and closet bolts.

Fig. 12-2. Closet flange with double wax rings installed.

Fig. 12-3. Back-to-back comparision of a standard wax ring and a wax ring with a plastic horn.

Lavatories

When the inspector checks the lavatory installation, he will look for many defects. He will look to see if you have created a watertight seal between the lavatory bowl and the countertop. If you haven't, run a bead of caulking around the lip of the bowl to seal it. Lift the pop-up rod to seal the drain of the bowl. Fill the bowl with water until it is just below the overflow hole. While the bowl is filling, look at the stream of water coming from the faucet's spout.

The stream should come out evenly. If it sprays in various directions, the aerator needs to be cleaned or replaced. If necessary, unscrew the aerator and inspect it for debris. It is not unusual for aerators to become clogged with flux, mineral deposits, and other objects when the water to the faucet is turned on after a new installation. Clean or replace the aerator to ensure a smooth stream of water from the faucet.

As you fill the bowl, check to see that the hot and cold water is piped correctly to the faucet. The hot should be piped to the left side of the faucet, and the cold should be piped to the right side. If the pipes have been crossed during the rough-in, install new supply tubes to correct the problem. Simply cross the supply tubes beneath the lavatory to get the hot and cold water supply to the proper faucet inlets.

With Moen single-handle faucets, you may turn the cartridge to reverse the hot and cold. To do this, cut all the water off to the faucet. Remove the plastic cover that hides the screw in the faucet's handle. Remove the screw and handle from the faucet. With a pair of needle-nose pliers, grasp the clip that holds the cartridge in place. Pull the clip out and turn the cartridge 180 degrees. Replace the clip, handle, screw, and cover cap. The hot and cold water now should be on the proper sides.

When the lavatory bowl is full of water, depress the pop-up rod and drain the bowl. Wipe all the slip nuts on the drainage and trap to check for leaks. If you find a leak, tighten the slip nuts. While you are under the lavatory, wipe down all the water supply connections. When you are sure you don't have any leaks, move on to the bathtub or shower.

Bathtubs and showers

As long as the shower base drains freely, there is nothing else you need to test in the drainage. Turn on the water and confirm that the hot and cold is piped to the proper inlets. If the water pipes are crossed, you will have to gain access to the shower valve and re-pipe the supplies to the valve. There is no easy way out of this one, unless the shower valve is a Moen single-handle unit. With a Moen single-handle valve, you may turn the cartridge in the manner described for lavatory faucets.

The last thing to inspect on the shower is the shower head. Check the connection between the shower head and the shower arm for leaks. If there is a leak, tighten the connection. If it continues to leak, remove the shower head, apply new pipe dope, and reinstall the head. If the spray from the shower head is not even, remove the head and inspect it for foreign objects blocking the waterways. The waterways of a shower head are susceptible to the same blockages as an aerator.

Bathtubs require a few additional checks. In addition to the inspections performed on a shower, you will have to test the diverter of the faucet if you have a combination tub and shower. The diverter is often on the tub spout, but it might be the center handle of a three-handle faucet or a small button in the escutcheon of a single-handle faucet. With the water on and coming out of the tub spout, engage the diverter. Water should come from the shower head instead of the tub spout.

If water continues to run strongly from the tub spout, repair the diverter. You may have to clear the diverter of foreign objects or replace a washer. Refer to the owner's instructions for exact information on how to correct the diverter's defect.

You will also have to test the drainage mechanism of the tub. Operate the tub waste and seal the tub's drain. Fill the tub with water and watch to see that it retains the water. If the water leaks out, adjust the tub waste to correct the problem. Operate the tub waste and release the water. See that it flows down the drain smoothly. If the water drains slowly, adjust the tub waste to correct the problem.

Kitchen plumbing

The tests performed on the kitchen sink will be similar to those done on a lavatory. Check for caulking around the sink and to see that the hot and cold water is piped properly. Check the aerator and drainage fittings. To test the drainage, fill all sink bowls to capacity before you empty them. By filling the sink bowls, you increase the pressure on the drainage connections. This pressure will reveal leaks that a stream of running water will not. The inspector will test the sink by filling the bowls. Be sure you do, too. Check the cut-offs and supply tubes for leaks.

If you have a garbage disposer, fill the sink bowl with water and release it all at once. Cut the disposer on while the water is draining. Inspect all the fittings on the disposer for leaks. If you have a dishwasher, run it through a full cycle to test the plumbing. This will allow you to test the water and drainage connections associated with the dishwasher. In the case of saddle valves, be sure the valve is open and not leaking.

Water heaters

There isn't much to test on water heaters. If you have installed the heater to code, all you have to test is the relief valve. Lift the handle of the relief valve to allow water to run through the discharge tube. As long as the water runs when the handle is up, and stops when the handle is down, you should be all set.

Hose bibs

The inspector will look to see if you have secured the hose bibs properly. You should have the hose bib screwed to the surface behind it. If the hose bib is secure, test to see that cold water comes out of the valve. As long as the cold water cuts on and off properly, you shouldn't have a problem with the official inspection.

Water pumps

If you have a water pump, inspect all connections for leaks. Confirm the water pressure in the storage tank. If the system was piped to code, these will be the only items inspected.

Laundry hookups

In the final inspection, the washer hookup will be tested for leaks and to confirm that the hot and cold water is piped to the proper place.

Miscellaneous plumbing

Laundry tubs and bar sinks will be tested like kitchen sinks and lavatories. Miscellaneous plumbing will be scrutinized for the same installation procedures described for the above fixtures. Essentially, check all the fixtures to be sure they work properly.

Closing comments

By taking the time to perform your own final inspection, before you call the inspector, you can reduce the risks of a rejection slip. Having the final plumbing rejected will delay the job and will most likely cost you a re-inspection fee. Making a thorough inspection of your own can save you time, money, and embarrassment.

Part III

Remodeling jobs, from planning to inspection

13

Planning
a remodeling job

UP TO NOW, we have talked mostly about plumbing for new houses and new construction. The following chapters will take you into the often challenging world of remodeling. Remodeling places demands on a plumber that are never encountered with new-construction plumbing.

The principles of plumbing are the same for both new and old work, but that is frequently where the comparison ends. Existing conditions and old plumbing will have a major impact on what you are able to do with the new plumbing. When it comes to installing plumbing in a remodeling job, you will have to spend extra time in the planning stage.

Design considerations

The first logical step to any plumbing installation is to draw a working design. This process was covered in Chapter 3 for new-construction plumbing. When you are working with a remodeling job, you will use most of the advice in Chapter 3, but you will need to expand on the information. This chapter will help you to look ahead and identify the differences you may expect when working with old plumbing. Chapter 14 will address potential pitfalls and problems. Your design will be the most effective when you combine the knowledge gained from these two chapters.

When you plumb a new house, you are responsible for every aspect of the installation. You can control the type of materials you will use and how they will be installed. With enough planning, you can design a plumbing layout that will allow you to place fixtures in the locations of your choice. This is not the case with the plumbing for remodeling work.

When you must make the new plumbing meld with existing pipes, you have limited options available. The seemingly simple act of replacing existing fixtures can turn into a major job. The plumbing code allows some latitude for plumbers combining new plumbing with old plumbing. While these code exceptions help, they do not solve all the hurdles you may be required to clear.

When you begin the plumbing design for a remodeling job, make sure you have plenty of paper on hand. The design is likely to require many changes. In new work, you can often work up the design from a set of blueprints. Don't even consider designing a system for a remodeling job without an on-site visit.

There is no way you can work a remodeling job with only a set of blueprints. It is essential that you complete a comprehensive inspection of all existing conditions before you design the plumbing system. In most cases, you will have to make multiple visits to the job before you can develop a successful design. If you happen to live in the house where the work will be done, you have an advantage over an outside plumber. As the resident, you can clarify your questions on the design as you draw it.

Identifying the type of job

The plumbing for remodeling work can take many forms. Before you can design the system, you must classify the type of work required. The easiest type of remodeling work is installing plumbing for a new addition to the home. This type of work will be referred to as new-addition plumbing. Replacing existing fixtures will not require much of a design, but the task might become complicated. This type of work will be called fixture replacement. Adding new plumbing within the existing structure is usually the most complicated type of remodeling. This work will be identified as add-on plumbing. The last classification will be pipe replacement. This category will cover the replacement of drain-waste-vent (DWV) and water distribution systems.

Designing pipe replacement

Developing a design for pipe replacement may seem simple, but it will require more thought than you think. When most people consider replacing existing pipe, they assume it will be just a matter of installing new pipes where the old ones were located. This assumption might lead you into a trap. Typically, when existing plumbing needs to be replaced, the system is outdated. Most plumbing codes require the plumber to bring the old plumbing up to current code requirements if a substantial amount of the system is replaced. This fact can transform an apparently easy job into a rigorous task.

Before embarking on extensive pipe replacement, consult your local plumbing inspector. If you have to bring the old system up to present code, the extent of the work may grow significantly. You may have to run larger pipes or add vents. The details of these complications will be discussed in the next chapter.

Making the design for a simple replacement is uncomplicated. If all you have to do is make a direct replacement, draw the design by copying the existing piping scheme. If you must add vents, you will have to draw a design indicating the sizes and locations of the vents. On paper, this design will be drawn like the designs for new-construction plumbing. On the job, the design may have to undergo several changes to avoid existing obstacles.

The most effective way to create a sound design for major changes is to sketch the design while you are at the job site. Then, as you draw the schematic, you can spot some of the problems around which you will have to work. If you are going to have to run pipes in existing walls, you may want to hold off on the final design until the walls have been opened.

Go over the design with the remodeling contractor. Chances are the contractor will find flaws in the design. After meeting with the contractor, make any necessary design changes. If you will not be able to postpone the final draft until all the walls are opened, go ahead and apply for the permit. Even with the best design, expect to make additional changes once the work is started.

Fixture replacement

You will not need a fancy design for fixture replacement, but you will need to do some investigative work. Look over the existing fixtures and note key elements in their placement. Permits are not usually required for fixture replacement, unless you are relocating the fixture. Even though you don't need a technical design, you will need to pay attention to detail.

New-addition plumbing

New-addition plumbing is probably the easiest type of remodeling work because it allows the most freedom. The plumbing will go in much the same as new-construction plumbing. Likewise, the design will be drawn with methods similar to those used in new construction.

You will need to inspect the existing plumbing and structural conditions before you can complete the plumbing design. Once the new plumbing reaches the foundation of the new addition, you will have to connect to the existing plumbing. When you draw the design, determining how you will connect to the existing system might get tricky.

Will the new drainage system connect to the building drain of the existing home, or will it have to tie into the sewer outside of the home? This decision will have to be made before you can complete the plumbing design. Not only will you need to know how the connection will be made, the code enforcement officer will want to know how you plan to make the connection. Before a plumbing permit will be issued, you must know where the new drainage will meet the old plumbing.

Designing a path for the water distribution system that will serve the addition might cause you some trouble. Will the existing water service handle the increased load of fixture units from the new plumbing? Is the existing water heater adequate for the new plumbing? How will you get the new plumbing to the location of the existing water pipes? All of these questions must be answered before you develop the design. The plumbing inspector will expect you to show clearly how the merger between the old and the new will be made.

Add-on plumbing

Add-on plumbing offers the greatest challenge when it comes to drawing an accurate plumbing design. Converting an attic into living space is a typical add-on plumbing situation. If a bathroom will be added in the attic, you must figure out how to get the pipes to the new bathroom. Finding a suitable path up through the existing walls may be taxing on all but the most experienced plumbers. Trying to work a 3-inch drain up through finished living space, without destroying the home, is no easy task.

Coming up with a dependable design for this type of plumbing can keep you up nights. A site visit is mandatory for add-on plumbing. Depending upon the amount of cooperation you receive from the general contractor, add-on plumbing may seem impossible to design. As you attempt to find a suitable direction for the pipes, you will be constantly plagued by barriers.

There is no clear-cut way to design an error-free plan for installing the plumbing for an add-on system. The best you can do is design a system you think will work. Until existing walls are opened, you will not be certain of the effectiveness of the design.

Often, the starting and ending points of the system may be the only definite factor you know when you lay out the fixture design. Use this information to put the fixture locations into perspective with the pipes to which you will connect. Once you know where the drains and water pipes will connect to the existing plumbing, you can make a rough draft of the overall design.

You will have to make numerous measurements in order to evaluate the best places to install your new plumbing. When you design the plumbing for an add-on job, it is best to go over the job description with the general contractor. Find out what walls will be opened or removed. Ask about any new hindrances that may impede the installation of the system. Will steel be added to support the additional weight when the attic is converted to living space? The remodeling contractor should be a great source of help as you plot a successful course for the plumbing.

Fixture locations

Choosing fixture locations for remodeling jobs can be a challenge. Fixture location dictates the method of the basic plumbing installation. Unfortunately, in remodeling jobs existing conditions often determine fixture locations. This chain of events can alter the plan considerably.

When you are designing the plumbing for a remodeling job, you must be aware of concealed obstacles. The obstacles can take many shapes. Fire blocking in walls may prevent the installation of stacks and risers without the cutting of finished walls. In prime conditions, you may be able to work these pipes up from a basement to an attic without affecting the finished walls, but you can't count on it. If the walls contain fire blocking, you will have to open them to get the pipes up into the attic. If you encounter fire blocking, your only option, other than cutting into existing walls, is to construct a pipe chase.

Electrical wires can present the same type of problems as fire blocking. If the wiring is scattered through existing walls, you may not be able to snake the pipes up into the attic. Heat ducts and existing plumbing provide additional risks. You must be prepared to develop a contingency plan when you install new plumbing in existing walls.

Some old homes do not have conventional framing members in the structural system. When you try to cut the hole for a shower or a closet flange, you could hit a log beam. Old houses frequently have logs, sawn in half, for floor joists. A simple header between the joists will not get the job done under these conditions. Evaluate the age of the house you are working with and expect the unexpected.

If you come into contact with plaster walls, your game plan may have to change. The plaster and lathe may restrict the size of the pipes that may be installed in the wall cavity. If you are counting on getting a 3-inch pipe into the wall, you may be out of luck with a plaster wall. Cutting and repairing plaster is more expensive than working with drywall. If you are the one paying for the repair of these walls, there will be extra money coming out of your pocket.

Planning pointers

Another often overlooked question is that of existing plumbing. As you tour the job site, make note of the size of the existing pipes and inspect the condition of the existing pipes to which you will connect new pipe.

The sewer

Check the size of the sewer pipe where it enters the home. If it is a 3-inch pipe, you could be in for big trouble. A 3-inch sewer will not carry the discharge of more than two water closets. If you are planning to install a third water closet on a 3-inch sewer, you are out of luck. Before the job can be done, you will have to install a new sewer. It may be more cost effective to run a new sewer, independent of the old sewer, for the new plumbing. In most cases, it will make more sense to replace the existing sewer with a larger one. Before you become too involved in designing the remodeling job, make sure the sewer is large enough to carry the increased fixture-unit load.

The water service

As with the sewer, check the size of the incoming water service. Verify that the water service is able to provide an adequate water supply. Only in rare circumstances will you have to replace the existing water service with a larger one.

The water distribution system

These pipes must be inspected to assure adequate size. Take note of the type of pipe used in the water distribution system since you will have to connect the new plumbing to these existing pipes.

A common problem with older homes is the size of the water pipes, especially the hot-water pipes. It is not unusual to find homes where all the hot-water distribution is made through 1/2-inch pipes. As you know by now, 1/2-inch pipe is not large enough to supply the hot-water for a house. If you can't make connections to 3/4-inch water pipes, you may have to apply for a variance. Generally, if the only water pipes in the house are 1/2-inch, the code enforcement office will allow you to connect to them for the new installation.

While this is a common practice, don't count on it. The code office may require that you run the proper size piping for the installation. If this happens, they may also make you bring the remainder of the home's plumbing up to current code requirements. If at any time you have questions about your responsibilities to the plumbing code, consult your local plumbing inspector.

Old plumbing that must be moved

It is easy to get caught up when figuring out how you will install the new plumbing, and not consider what may need to be moved in the existing system. If an attic is being converted to living space, the existing vents may have to be relocated. If existing walls will be removed, there may be concealed plumbing to move. It may be water pipes, drains, or vents. This is a potential task that may be difficult to foresee.

When discussing the overall remodeling plans with the general contractor or owner, take wall removal into consideration. If a wall is going to be removed, any pipes in the wall will have to be relocated. When a wall is scheduled for demolition, look in the attic and in the basement or crawl space for pipes entering or exiting the wall. If you see pipes that will have to be moved, take note of the type of pipe you will be moving. Determine the pipe material used and its purpose. This information will help you design a plan for the successful relocation of the pipe.

After you have an initial design on paper, go back through the job and look for problems you missed on the first walk-through inspection. When possible, discuss your plumbing plans with the general contractor and any subcontractors involved in the job. If the heating-ventilation-air conditioning (hvac) contractor has chosen the corner of a closet for a duct chase, you will be unable to use the space for the plumbing. If you don't own the property, talk your plans over with the owner. When all reasonable steps have been taken to avoid conflicts, make the final draft of the plan. Use the final plan to apply for the plumbing permit.

When remodeling, the best laid plans often go astray. While drawing a working design may seem futile, it is necessary to obtain the permit. Anyone in the business knows there is a high probability that changes will be made to the remodeling plans. Experienced inspectors will anticipate on-the-job changes. As long as the changes comply with code requirements, they will not hold you to the initial plan. If you encounter an impossible situation, discuss it with the code officer. In extreme cases, the local authority has the power to grant a variance from the plumbing code.

Variances are not given indiscriminately. The fact that your job will be more difficult when made to comply with code regulations is not reason enough for a variance. Don't ask for or expect a variance until you have exhausted all reasonable efforts to meet code requirements. When you have no choice, ask the inspector to visit the job with you. Go over your dilemma and ask for advice. Never start the negotiation for a variance by telling the code officer how you are going to do the job. Always ask for his opinion on the best solution to the problem. If the inspector draws a blank, then you can offer your suggestions for maneuvering around the obstacle.

Chapter 14 discusses many of the problems and pitfalls you may experience during the installation of the plumbing for these jobs. Before you attempt to draw the design, read the next chapter. By combining the information in Chapter 14 with this information, you will be prepared to make a working plan for your remodeling endeavors. Chapter 14 will also help you to overcome on-the-job obstacles. By paying close attention to the examples in Chapter 14, you can avoid the problems most often found in remodeling work.

The benefits of preliminary planning

Preliminary planning will help you in a host of ways. It will eliminate some problems. The problems that are not removed will be easier to overcome with strong planning. You will spend less time running back and forth after materials when you have a feasible plan. The chances of damaging the walls, ceilings, and other existing job conditions will be reduced by proper planning. Your work will go faster, saving you time and money. While the planning efforts for remodeling work may seem dismal, they will pay dividends in the end.

14

Remodeling pitfalls and problems

WHEN YOU ARE installing plumbing for a remodeling job, you should expect unforeseen problems. Unless you have years of remodeling experience, plumbing a remodeling job will test your mettle. Planning is crucial for this type of plumbing, but the best planning may not predict all the pitfalls. The factor that does the most to remove potential pitfalls and problems is experience. For you to gain field experience, you will be a victim of circumstances and mistakes.

This chapter is written to help you avoid the pain of learning from on-the-job experience. Don't get the wrong idea, you will still experience your share of on-site surprises. But, if you pay attention to the advice in this chapter, you might eliminate many of the most common faults. As a master plumber, I have worked with remodeling jobs for more than 17 years. This chapter is a compilation of my 17 years of in-field experience.

The water heater

With average remodeling jobs, the existing water heater will be sufficient. However, if you are installing a whirlpool tub or anything that will demand large amounts of hot water, consider the possibility of adding a new water heater.

Pipe problems

When you remodel and replace or add on to an existing plumbing system you will probably encounter problems with the existing pipes. I discuss a few problems, cautions, and solutions below.

Heat pipes

If you are remodeling an old house, you may find heating pipes mixed in with water pipes. Older homes are frequently equipped with steam or hot-water heating pipes. It may be hard to tell these pipes from water pipes, unless you are paying close attention to the job. I have had plumbers cut into heating pipes, thinking they were potable water pipes, and pipe out an entire plumbing system from the wrong pipes. This is not only embarrassing, it causes extra work and lost money.

Steam pipes are not usually mistaken for water pipes because of their large size. However, forced hot-water heat pipes are the same size as many water lines. In the winter, the pipes usually can be distinguished by their high temperature. If you lay your hand on an active pipe, it will not take long before you realize it is a heating pipe. But in warm weather, the heating pipes may not be in use and there is more of a chance for a mistake.

Rather than assume you are connecting to water pipes, trace the pipes to be sure. In most cases, you will need to identify the hot and cold water pipes. If you don't verify the nature of the pipes, you could contaminate the potable water pipes with ugly water from a boiler. It only takes a few minutes to establish the identity of the pipes. It takes hours to correct a mistake with heating pipes.

Galvanized water pipes. If you find the existing water distribution system to be piped with galvanized steel pipe, give serious consideration to replacing all of the galvanized pipe. If problems are not present now, they will arise in the future. Galvanized steel pipe is famous for rusting from the inside out. As the pipe rusts, two conditions develop that affect the water distribution system.

The first condition is restricted water flow within the pipe. As the pipe rusts, the rough surface collects minerals and other undesirable objects that restrict water flow. In time, the pipe can become closed to the point where only a dribble of water will come out of the faucets.

The second effect of rusting galvanized water pipe is the development of leaks. Where the pipe has been threaded, the wall of the pipe is thinner. The threading process weakens the pipe and rust attacks these weak spots. As the rust continues, the pipe threads deteriorate and leak. Replace all the galvanized water pipe, unless there are extenuating circumstances.

If you are working with galvanized or brass water pipe, you will have to be careful of the existing joints. By the time you begin remodeling, these threaded joints may be ready to break. As you work with the pipe, your actions may cause the joints to break loose at any point.

The vibration from a reciprocating saw is enough to cause leaks in weak joints. Twisting and turning the pipes with a pipe wrench is more than enough to cause trouble. As the pipe is stressed, joints several feet away may crack or break. Often, these leaks will not be discovered until the water is turned back on. By the time you react to the leak, significant water damage can be done.

When working with old, threaded pipe, be as gentle as possible. Avoid putting undue stress on old joints. When you cut on the water to test for leaks in the new work, inspect all the old piping. Look closely for leaks. A cracked thread may only produce a small leak, but it can do big damage over time.

Illegal piping. On numerous occasions I have found existing plumbing systems that were piped with illegal materials. The most common violation in the water distribution system is the use of polyethylene pipe for the entire water distribution system. Since this pipe is not approved for use with hot water, it cannot be used as a water distribution pipe in houses with hot water. The cold water pipes must be made from the same material as the hot water pipes.

The most outrageous illegal installation I have ever seen in a drainage system involved the use of slotted pipe for a building drain and sewer. Slotted pipe is used in septic fields. The bottom of the pipe is solid and the top of the pipe is slotted with holes. These holes allow the effluent carried in the pipe to seep out of the pipe and into a septic field. This seeping action is fine when the pipe is used for a septic system. It is not acceptable when the pipe is used for a building drain or sewer.

I went to the job to give the homeowner an estimate for adding a new bathroom. When I crawled under the house to assess the existing plumbing, I could hardly believe what I saw. All of the 3-inch drains were piped with slotted pipe. The ground under the toilet was saturated with raw sewage. As the toilet was flushed, the water pressure forced sewage to escape from the slots in the pipe. Not only was this disgusting and illegal, it was potentially dangerous.

When the sewer was dug up between the home and the septic tank, the ground was holding raw sewage. When the sewer was installed, it was not properly graded. Waste from the house was building up in the pipe and spilling over into the ground. This example is an extreme case, but beware of illegal piping arrangements that will affect the new plumbing installation.

Galvanized steel drains. Galvanized drains have earned quite a reputation in the world of pipe blockages (Fig. 14-1). Like galvanized water pipes, galvanized drains fall victim to rust. In the case of drains, the inside roughness, caused by rust, catches hair, grease, and other unidentified cruddy objects. As these items become snagged on the rough spots, the interior of the drain slowly closes. In time, the buildup will block entirely the passageway of the pipe. The result is a drain pipe that will not drain.

When this type of blockage is attacked with the average snake, the snake only punches a hole in the blockage. The drain will work for awhile, but it will not be long before the pipe is stopped up again. Unless the blockage is removed with a cutting head on an electric drain cleaner, the pipe will become restricted soon after the snaking (Fig. 14-2).

When a garbage disposer is added to a kitchen sink with a galvanized drain, expect problems. Since garbage disposers send food particles down the drain, the rough spots in the drain catch and hold the food. You should count on having to replace any old galvanized pipe used as a drain for garbage disposers. In addition to the rust problem, the fittings used with galvanized pipe are not as effective as modern fittings. New fittings utilize longer turns than those used with galvanized pipe. A galvanized quarter-bend takes a much sharper turn than a schedule 40 plastic quarter-bend. These tight turns are responsible for inducing stoppages within the pipe.

Fig. 14-1. Clogged galvanized drain pipe.

Fig. 14-2. Large electric drain cleaner.

In addition to habitual stoppages, galvanized drains will develop leaks at their threads. These are caused when the rust works on the weak spots at the threads. In general, galvanized pipe should be replaced whenever feasible.

Cutting cast-iron pipe. When remodeling, it is not unusual to be working with cast-iron drains and vents. Cutting these pipes to install a new fitting can be very dangerous. Many of these vertical pipes are not secured well. When you cut the pipe, the pipe above you may come crashing down. This is a potentially fatal situation.

If you must cut into a vertical, cast-iron stack, be sure it will not fall when it is cut. Find a hub on the pipe and use perforated strap to secure the pipe. Wrap the galvanized strapping around the pipe, under the hub, and nail it to the studs or floor joist. Whenever possible, have a helper on hand when you cut the stack. The method used to secure the pipe will vary with individual circumstances, but be sure the pipe will not move when it is cut.

Cast-iron pipe may be cut with roller cutters, metal-cutting saw blades, and some people can even cut it with a hammer and chisel. For the average job, rachet-style snap cutters are the best tool for cutting cast-iron pipe. These tools may be used to cut pipe when there is very little working room. If you can get the cutting chain around the pipe, and have a little room to work the handle back and forth, you can cut the pipe quickly and easily. These soil pipe cutters are expensive, but you can rent them from most rental stores.

If you can not use rachet-style soil pipe cutters, your next best option is a reciprocating saw with a high-quality, metal-cutting blade. It may take up to 20 minutes for each cut, but the job will get done. Whenever possible, use the rachet-style soil pipe cutters. These cutting methods apply to both types of cast-iron pipe. Remember, cast-iron pipe is heavy. When you cut the pipe, be sure the pipe is supported so that it doesn't fall on you.

There are times when the cast-iron pipe you are cutting is soft. When the pipe has gotten soft, or rotted as it is called in the trade, it does not cut well. Soil pipe cutters work on a cutting-wheel and pressure principle. As you apply pressure, generally by cranking up and down on a handle, the cutting wheels cut the pipe with a snap. This works fine for most cast-iron pipes, but not for soft pipe. When the pipe has rotted, the cutters will not make a clean cut. Instead of a crisp snap, you get a soggy crunch. When this happens, you have to employ other means to remove the section of pipe.

Many inexperienced plumbers create big trouble for themselves under these conditions. If they have worked around a master plumber, they have seen him handle this situation. Thinking that they can do the job, they whack the cut section of pipe with a hammer, only to see chunks of cast iron go down the drain. When the cast iron breaks into pieces and falls down the drain, these poor plumbers have a big mess on their hands.

The jagged pieces of pipe will lodge in a fitting and create a stoppage in the drainage system. A snake is useless when you try to remove such an obstruction. The drain will have to be opened and the pieces of pipe lodged in it removed. Getting to the blockage often requires breaking up a concrete floor or doing extensive excavation work. You can imagine the time and expense involved in correcting such a problem.

A master plumber will rarely create this type of situation. When the cast-iron pipe is rotten, a master will use different tactics to remove it. The initial cuts will be made with the soil pipe cutters, but from there the master will use skillful experience to avoid a bad situation. Instead of pounding on the pipe with a hammer, a master will use a hammer and a chisel. The chisel allows the pressure of each blow to be directed to the cut in the pipe.

At worst, a master will have a few tiny pieces of pipe escape into the drainage system, but they will not be the large, unmovable objects described earlier. Once the cutline has been defined with the hammer and chisel, an experienced plumber can tap out the cut section of the pipe. The process takes some time, but it does not create the disaster of the entire drainage system being blocked.

When there is so much pressure on the pipe that the section cannot be tapped out, the master will remove it in small pieces. He will create a small hole in the section with his hammer and chisel. Then he will insert the jaws of a pipe wrench into the hole. He will use a pipe wrench as a lever to extract small sections of the pipe. The leverage offered by the pipe wrench is more than a match for rotted cast-iron pipe. After nipping away for awhile, there will be a slit in the pipe with which the plumber will work.

Once there is a sufficient opening, the plumber can use his hammer and chisel to drive out the section of pipe. The chisel is inserted through the slit. This allows all of the pressure from the hammer to be exerted on the inside of the rotted pipe. When the pipe breaks, it breaks outward and falls on the floor. This prevents large sections of the pipe from falling into the drainage system.

These removal methods are perfected with experience. For your endeavors, let me give you some advice. If the pipe does not snap cleanly, don't indiscriminately bang the cut section with a hammer. If you want, try to follow the instructions given above. If you find them difficult, call a professional. You may pay extra to have the section removed, but the money spent will be less than the cost to remove large pieces of cast-iron pipe from the drainage system.

In many cases, gently tapping the pipe on the cut will work. Avoid direct hits on the center of the pipe. These direct hits will cause the pipe to crack and fall. If you methodically tap the pipe, from an angle there is a good chance you will get it out without irreparable damage.

How far do you go?

When you evaluate the removal of existing plumbing, you must determine where you will end the removal process. As a rule-of-thumb, replace as much of the undesirable pipe as reasonable. Time, money, and concealment all play a part in your decision. If you leave questionable pipe in the system, you may regret it later. It is much easier to replace the old plumbing while you are in the middle of remodeling than it will be six months after the job is finished.

Can you imagine your anger when the new ceiling and carpet is ruined by water leaking from an old galvanized tub drain in the ceiling? How will you feel when you turn on the new shower and the water barely drips out of the shower head because of old galvanized water pipes that have closed? Evaluate the circumstances and make your own decision, but I strongly suggest you replace as much of the old piping as you can.

Inadequate septic systems

If you are adding plumbing to a house that uses a septic system, be careful. The septic system may not be of an adequate size to handle the increased demand from new plumbing. Before you add plumbing to an old septic system, have the system checked. You may be able to determine the system's capability by reviewing the original installation permit and plans. If this paperwork is not available, hire a professional to render an opinion on the ability of the system to handle the increased load.

Septic systems are expensive to install and expand. It would be a shame to add a new bathroom, only to find that the septic system is not adequate to process the waste. The cost of adding new chambers or lines in the septic field might shock you. While you may not be happy to discover you need to invest extra money in the septic system, it will be better to find it out before the new plumbing is installed.

Rotted walls and floors

When you plan to replace old plumbing fixtures, you can never be sure what you will find during the process. It is not uncommon for toilets to leak and rot the floor beneath them. Bathing units are subject to the risks of rotten floors and walls. These structural problems will complicate your life as a plumber.

When you remove the old toilet and find a rotten floor, someone will have to pay to have the floor repaired. If you are the homeowner, that someone will be you. Toilets contribute to the rotting of floors through leaks around wax seals, tank-to-bowl bolts, and condensation. While the quantity of water running onto the floor may be minor, the damage can be major. As time passes, the water works its way into the underlayment, sub-floor, and floor joists. Small leaks may not stain the ceiling below a toilet until the damage is done.

If you are concerned about water damage around the base of the toilet, you can test for it before removing the old toilet. Take a knife or screwdriver and probe the floor around the base of the toilet (Fig. 14-3). If the point of the probing instrument sinks into the floor, you have a problem. If the water damage is only in the underlayment or sub-floor, the cost of repairing the damage will be minimal. If you were not planning to replace the floor covering, you will incur additional expense. To correct the problem, the finished floor covering will have to be replaced. Trying to make a patch in the floor covering will result in a floor that does not match.

Bathtubs and showers may have caused hidden damage to the floor or walls. When people step out of the bathing unit, they often drip water on the floor. If there is not a good seal at the base of the tub or shower, this water will run under the bathing unit. Prolonged use under these conditions will result in severe water damage to the floor and floor joists.

If the walls surrounding the bathing unit leak, water will invade the wall cav-

Fig. 14-3. Probing the base of a toilet to check for bad floor conditions.

ity. The water may rot the studs or the base plate of the wall. Given enough time, the water will penetrate the wall's plate and enter the floor structure. This type of damage may go unnoticed until you remove the old fixture. When the damage is found, it will have to be repaired before you may move ahead with the plumbing.

The structural damage caused by water can be substantial. I have seen floors rot to the point where the fixture falls through the floor. It is not only embarrassing to have the toilet or tub fall through the floor; it is dangerous. If you will be responsible for the costs incurred to repair water damage, allow for this possibility in your remodeling budget.

Living creatures

Depending on where you live, you may confront any number of various living creatures. A partial list of these living obstacles includes: bats, snakes, rats, bees, and skunks. During my career, I have run into all of these creatures, and more. Attics and crawl spaces are the most common places to find these critters, but they also may be in the walls of the home.

I have cut open walls and been attacked by swarms of bees. Going into attics to run vents, I have shared the space with bats, squirrels, and raccoons. Crawling under homes, I have come face to face with rattlesnakes, rats, feral cats, and other things that make the hair on the back of your neck stand up. While there is little you can do to ensure you will not have a run-in with wild animals, caution is your best defense.

Crooked walls

With old houses, and some not-so-old homes, the walls are out of plumb. These crooked walls can give you a fit when you set the fixtures. Bathing units are particularly susceptible to this problem. Before you install the new plumbing, check all walls to be sure they are plumb. If the walls are not plumb, have the carpenters correct the problem while the walls are open. If the wall is not straightened out before it is finished, you will have to make up for the crooked walls with caulking and jury-rig installation methods.

Cut-offs that won't

Remodeling jobs frequently are complicated by valves that will not close. In these cases, you may have to cut the water off at the water meter. Never assume that an existing valve will function properly. Cut off the valve and confirm that the water is off by opening a faucet. Many old valves become stuck and will not close. Even when the handle turns, the valve may not close. If you fail to double check the effectiveness of the valve, you may flood the home when you cut pipes or loosen connections.

When you have a valve that is not cutting off the flow of water, you will have to take other steps. Usually, the easiest solution is to cut off the valve that serves the entire water distribution system. If this happens to be the valve that is not working, you may have to take your actions a step further. If the water comes from a private well, cut off the electrical power to the pump. If the pump can't run, it cannot pump water. With the pump disabled, you can drain the water pipes and continue with your work. If the water comes from a municipal source, call the water department to cut off the water at the water meter. While you have the water cut off, replace the defective valve. The next time you need to cut off the water, you may not have time to wait for municipal employees.

Closing comments

Remodeling is filled with potential problems and pitfalls. Some of these setbacks cannot be avoided, but many can. By using the information in this chapter as a guide, you should be able to predict many of your job's potential problems. The key to success lies in planning and paying attention to details. Look ahead to eliminate problems before they occur.

15

Adapting and connecting to existing plumbing

ADAPTING NEW PLUMBING to an existing system may be perplexing for the inexperienced plumber when you consider how many approved plumbing materials exist. While the possibilities may seem endless, there is no reason for dismay. There are materials and methods available to make adaptations easy. This chapter will enlighten you as to the methods professionals use when working with a mix of materials.

Drain-waste-vent copper systems

There are still a large number of operating plumbing systems that were constructed from drain-waste-vent (DWV) copper. This thin-walled copper has a long life as a drain and vent material. When you find DWV copper, it is usually in good working condition. Since this pipe produces few problems, there is no need to replace it. However, it is rarely cost effective to use DWV copper to extend a system in today's plumbing practices. When you want to add to a system configured with copper, you will probably want to connect to the copper with plastic pipe.

When the goal is to mate plastic pipe with copper pipe, you have a few options. The most common methods for converting copper to plastic is the use of threaded connections or rubber couplings. Either of these adapters will give satisfactory results without much effort or expense.

Threaded connections with DWV copper

If you choose to make the connection between DWV copper and plastic pipe with a threaded connection, you will need soldering skills and a pretty good torch for soldering large pipes and fittings. There is a big difference between soldering a 3-inch copper drain and a $^1/_2$-inch copper water pipe. Additional heat is needed to make large soldering joints.

The small, hand-held torches used by many homeowners can make soldering drain pipes a drudgery. These small torches will let you accomplish the task with smaller DWV pipes, but you must make allowances for larger pipe sizes. When you solder DWV drains with a small torch, move the heat around the fitting. Heat the joint evenly and thoroughly. When the temperature is right, the solder will flow smoothly.

If you will be using 3- and 4-inch copper pipes, rent a larger torch. There are torch tips available that allow the pipe to be heated from two sides simultaneously. These yoke-type tips are horseshoe shaped and surround the pipe.

The first step when connecting to an existing DWV copper pipe is to cut the copper pipe. You will need to make room for the tee or wye. Roller-type cutters are the best choice for cutting DWV copper, if you have room to operate them. When space is limited, you may cut the copper pipe with a hacksaw or a reciprocating saw that has been fitted with metal-cutting blades. The principles for soldering DWV copper are the same as those used for copper water pipe.

Measuring for the cut and adding a new fitting

Before you cut the copper, you must consider how much pipe to remove for the fitting. Another consideration must be the flexibility of the existing pipe. If the existing pipe cannot be moved forward, backward, or vertically, you will have to use an additional fitting to make the connection. Once the pipe is cut, prepare and install the new fitting. It is important that all the pipes that enter the new fitting seat properly. This is not to say that the pipe must be jammed into the fitting to the hilt, but you must have enough pipe in the fitting to assure a solid joint. If the existing pipe has very little play in it, you may not be able to install the new fitting without a slip coupling.

Slip couplings do not have ridges near their center points like standard couplings. A slip coupling can slide down the length of a pipe without being stopped by the ridges found in normal couplings. This sliding ability makes a slip coupling indispensable for some adaptation work. To install a normal fitting, you must have enough room to move existing pipes back and forth. With slip couplings, this range of movement is not required.

If you are forced to use a slip coupling, cut additional pipe out of the existing system in order to install the new fitting. The new fitting will be installed on one end of the existing pipe. Slide the slip coupling onto the piece of existing pipe. Place a new piece of pipe in the vacant end of the new fitting. The new pipe should be long enough to reach nearly to the end of the existing pipe. Then slide the slip coupling back towards the new pipe. In doing so, you connect the two pipes without needing to move the existing pipes.

Installing adapters

Once the branch fitting is installed, you are ready to install the threaded adapter. The threaded adapter may have male or female threads. A piece of pipe will be installed between the branch fitting and the adapter. These joints will be soldered with normal practices. When the pipe and fittings have cooled, you are ready to switch pipe types.

When the copper has cooled, apply pipe dope to the threads of the male adapter. The male adapter may be copper or plastic, depending on the order you have chosen for the adapters. Screw the plastic adapter into or onto the copper adapter. When the plastic adapter is tight, the conversion is made. From this point, run the plastic pipe in the normal manner.

Rubber couplings

If you wish there was an easier way to convert DWV copper to schedule 40 plastic pipe, there is. Rubber couplings can make the conversion between the two pipe types simple (Fig. 15-1). When you use rubber couplings, there is no need for soldering and the amount of pipe you cut out of the existing system is not critical. The measurements are not critical, because you can bridge the span with the plastic pipe. Rubber couplings slide up the existing pipes, like a slip coupling, so pipes that won't move are no problem. All aspects considered, I can see no reason to use any other type of conversion method in normal applications.

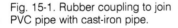
Fig. 15-1. Rubber coupling to join PVC pipe with cast-iron pipe.

The first step is to cut a section of pipe out of the existing plumbing system. Remove enough pipe to allow the installation of the branch fitting and two pieces of pipe. With the old pipe removed, slide a rubber coupling onto each end of the existing pipes. Hold the branch fitting in place and take measurements for the two pieces of pipe that will join the fitting to the existing pipes. Install the two pipes into the ends of the branch fitting.

Hold the branch fitting, and its pipe extensions, in place to check the measurements. The plastic pipes that protrude from the branch fitting should come close to the copper pipes. With the pipes in place, slide the rubber couplings over the plastic pipes. Position the stainless steel clamps in the proper locations and tighten the clamps. When the clamps are tight, the conversion is complete. Now you may work directly from the plastic branch fitting to complete the installation.

Galvanized drains

Many houses are still equipped with galvanized drains and vents. Due to galvanized pipe's potential for problems, it is a good idea to replace galvanized pipes whenever possible. On the occasions where you will not be replacing the pipe, you must use adapters to make the steel pipe compatible with new plumbing materials.

In many situations you will be forced to cut the old galvanized piping in order to install a branch fitting for the new plumbing. A hacksaw is the tool most often used to accomplish this task. If you have a reciprocating saw with a metal-cutting blade, you may use it. When you use an electric saw, be careful you don't get electrocuted. The saw must be insulated and designed for this type of work. Old drains often hold pockets of water. If you hit this water with a faulty saw, you will get a shock.

Rubber couplings

Rubber couplings are the easiest way to convert galvanized pipe to a different type of pipe. If you elect to use rubber couplings, the procedure will go about the same as described for DWV copper pipe. Cut out a section of the galvanized pipe to accommodate the new branch fitting. Then install the branch fitting and its pipe sections with the rubber couplings. The couplings simply slide over the ends of each pipe and are held in place by the stainless steel clamps.

Threaded adapters

Since you may not solder a fitting onto galvanized pipe, if you choose to use threaded adapters, you must have threads with which to work (Fig. 15-2). When you cut in a branch fitting, you must cut the old pipe where there is a threaded fitting or pipe end. This can be difficult. The old pipe may have concealed joints, or the pipe may be seized in its fittings. You will need pipe wrenches to disassemble old galvanized drains.

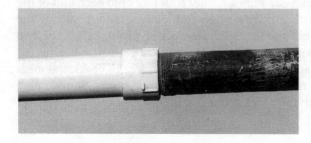

Fig. 15-2. PVC female adapter to join PVC pipe with galvanized steel pipe.

Once you have worked your way to a threaded fitting or pipe end, you can install the threaded adapters. Apply pipe dope to the male threads of the connection and screw the adapter into or onto the threads. Tighten the adapter with a pipe wrench. When the adapters are tight, proceed with the new piping material.

When you cut in a branch fitting on galvanized pipe, rubber couplings are the connections to make. The rubber couplings work fine and reduce the time and frustration involved in working with old galvanized pipes. If you are only converting a trap arm, or similar type of section, threaded adapters are a reasonable choice. With a trap arm, you will be unscrewing only one section of the steel pipe. Once it is removed, you will have the threads of the fitting on which to screw an adapter. Under these conditions, threaded adapters are the best choice.

Cast-iron pipe

Cast-iron pipe has been used for drains and vents for many years. Cast-iron pipe is still used for these purposes, though for the most part, cast-iron pipe has been replaced with plastic pipe in modern systems. There are two types of cast-iron pipe that you may encounter. The first type is a service weight pipe with hubs. This pipe is often called service weight cast-iron or bell and spigot cast-iron. The other type is a lighter weight pipe that does not have a hub.

Service weight cast-iron pipe is what you likely will encounter. It is the pipe used in most older homes. Service weight cast-iron pipe is usually joined with caulked lead joints at the hub connections. While caulked lead joints were used to install the old pipe, you may use modern adapters to avoid working with molten lead. The options for using adapters with bell and spigot cast-iron pipe will be discussed in the following paragraphs.

The lightweight, hubless cast-iron pipe used in some jobs is a much newer style of cast-iron pipe. The connections used for this type of pipe are a type of rubber coupling (Fig. 15-3). They are not the same rubber couplings referred to

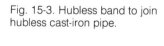
Fig. 15-3. Hubless band to join hubless cast-iron pipe.

throughout this chapter. These special couplings are designed to join hubless cast-iron pipes. The couplings have a rubber band that slides over the pipe and fitting. There is a stainless steel band that slides over the rubber coupling. The band is held in place by two clamps. When the clamps on the band are tight, they compress the rubber coupling and make the joint. These special couplings are easy to work with, but they are not your only option. The heavy rubber couplings discussed elsewhere in this chapter will work fine with this pipe.

Rubber couplings

The heavy rubber couplings, which have been discussed earlier, will work fine with either type of cast-iron pipe. Install them using the same method described earlier (Fig. 15-4). When working with hubless cast-iron pipe, you may use the special rubber couplings designed to work with the hubless pipe. If you use these special couplings to mate cast-iron pipe to plastic, you will need an adapter for the plastic pipe. It is possible to use the couplings without the plastic adapter, but you shouldn't. The plastic adapter will have one end formed to the proper size to accommodate the special coupling. Glue the adapter onto the plastic pipe and then slide the special coupling over the factory-formed end of the adapter.

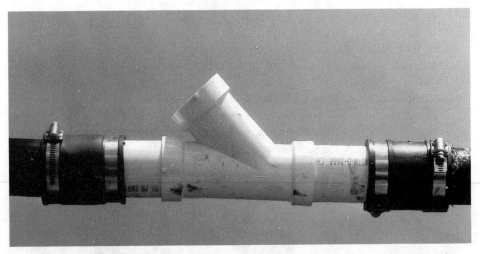

Fig. 15-4. Rubber couplings to insert PVC pipe and fitting between two pieces of cast-iron pipe.

Cast-iron doughnuts

When you work with the fittings for service weight cast-iron pipe, special adapters are available that allow plastic pipe to be installed into the hub (Fig. 15-5). Normally, when pipe is placed in the hub of service weight cast-iron pipe,

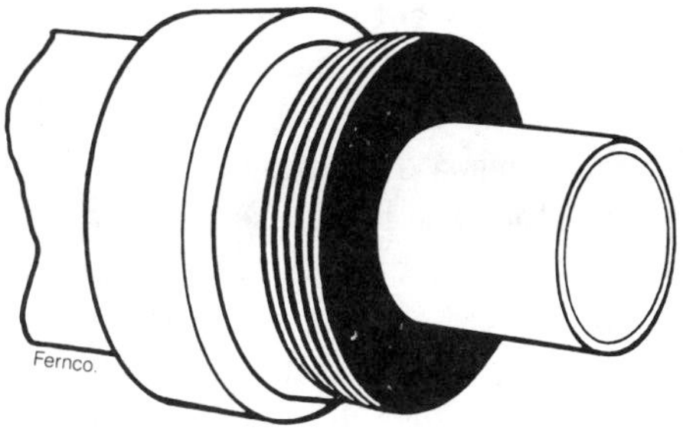

Fig. 15-5. Example of a cast-iron doughnut.

it is held in place by a hot lead joint. To avoid using molten lead, use a ring adapter. Plumbers call these ring adapters doughnuts.

These ring adapters are not always easy to use. They fit in the hub of the cast-iron pipe and the new pipe is driven into the ring. As the new pipe goes into the ring, the ring grips the pipe and forms a watertight joint. The problem comes when you try to get the new pipe into the ring. Generously lubricate the new pipe and the inside of the ring before you attempt the installation. Special lubricant is made for the job and sold by plumbing suppliers.

Place the doughnut in the hub of the fitting. Apply plenty of lubrication to the inside of the ring and the outside of the new pipe. Place the end of the pipe in the ring and push. The pipe will probably not go in very far. When the pipe is as far into the ring as you can push it by hand, you will have to use some force. Take a block of wood and place it over the exposed end of the pipe. Hit the wood with a hammer to apply even pressure to the pipe. With some luck and a little effort, you will drive the pipe into the doughnut. Once the pipe is all the way into the adapter ring, you may proceed with the new piping in the normal manner.

Copper water pipes

A vast majority of today's homes use copper pipe to distribute potable water. If you do not wish to make the new installation with the same type of pipe, you will have to use adapters. When you prefer to work with CPVC or some other type of plastic pipe, you must convert the copper to a suitable pipe type. This can be done with a variety of adapters. Of all the adapters available, threaded adapters are the most universal.

Painted copper pipes

The copper pipes in older homes may be covered in paint. At one time, it was fashionable to paint exposed pipes to blend in with the decor of the home. The paint may have made the pipes more attractive, but it definitely makes soldering these pipes more difficult. If you are required to cut a branch fitting into painted pipes, the paint must be removed from the area to be soldered.

Most plumbers cut the painted pipes first, and then they try to remove the paint. They usually use their regular sanding cloth and struggle to remove the paint. There is an easier way to remove the old paint. Remove the paint before you cut the pipes. Whole pipes are solid and easy to apply pressure to with the sandpaper. Once the pipes are cut, they wobble and bounce around, making it difficult to apply steady pressure with the sanding cloth. Another common problem when you cut the pipes before the paint is removed, is the water in the pipes. When the pipes are cut, water often runs down the pipe and gets the paint wet. The wet pipe gets the sanding cloth wet. With the pipe and the sandpaper wet, the paint is much harder to remove.

The standard sanding cloth used for copper pipe will remove most of the paint, but you will have to sand the pipe several times to accomplish the goal. If you know you will be working with painted pipes, put some steel wool in your tool box. Also invest in some sandpaper with a heavy grit. The rougher sandpaper will do a better job than the fine-grit paper usually used to sand copper. If you use the steel wool and coarse sandpaper before the pipes are cut, the paint should come off with minimal effort.

After the paint is removed, cut the pipe. Be sure the water is off before you cut the pipe. Drain as much water from the pipe as possible. If you can see water standing in the pipe, you know soldering will not be easy. It is not unusual for water to remain in pipes after they are cut. If the pipe runs horizontally, you may be able to pull down on the pipe to allow the water to escape. When you work with a vertical pipe, you will not be able to tip it down to drain it. Instead, you remove the water with a drinking straw. Don't worry, you will not have to suck the water out of the pipe. Place the straw in the vertical pipe. When the straw is in the pipe, blow through it. The pressure will force the standing water up and out of the pipe. Unless you have water leaking past a valve, this will clear enough water for you to do the soldering.

Apply a generous amount of flux to the pipe where the fitting will be placed. Before you install the fitting, heat the flux with the torch. As the flux bubbles and burns, it will clean the pipe. After this cleaning, follow the normal soldering procedures to complete the job.

Cutting copper pipe is not difficult. The easiest way is to use roller cutters, but you may also accomplish the job with a hacksaw. When you remodel, miniature roller cutters often come in handy. Measure to determine how much pipe you need to remove for a branch fitting using the same method described for DWV copper.

Threaded adapters

By using threaded adapters, you can mate copper water pipes with any other approved material. Cut in the tee and solder the threaded adapter to a piece of copper pipe at the point you wish to make the conversion. After the soldered joint cools, apply pipe dope to the male threads of the adapter to be used in the connection. Screw the male threads into the female threads until the fitting is tight. At this point, the conversion is complete. You may continue the new installation from the opposite end of the threaded adapter.

Compression fittings

Compression fittings are another possibility when you need to convert copper to a different type of pipe. Take a compression tee and cut it into the copper pipe (Fig. 15-6). The same measurement techniques used for copper fittings will work for compression fittings. Since the branch fitting is of the compression type, there is no soldering to be done.

Fig. 15-6. Compression tee.

Saddle valves

Self-piercing saddle valves are a good choice when you add new plumbing for an ice maker or a similar appliance. The saddle valve clamps to the copper pipe and allows you to use plastic or copper tubing between the saddle and the appliance (Fig. 15-7). The saddle valve connects to the tubing with a compression fitting and is held onto the copper water pipe with a clamp and bolt device. With a self-piercing saddle valve, you do not have to drill a hole in the copper water pipe. When the handle on the saddle is turned clockwise to its full extent, the copper water pipe is pierced. When the handle is turned back counterclockwise, the water will flow into the tubing.

Galvanized water pipes

Galvanized water pipes should be replaced whenever possible. If you prefer to connect new piping to the old galvanized pipe, you may use threaded adapters. The galvanized water pipe is cut using the same procedures described for galva-

Fig. 15-7. Saddle valve.

nized drains and vents. The threaded adapters are used in the same manner as those on the drains. The only difference is the size of the pipe and fittings.

CPVC water pipes

The most common way to adapt CPVC water pipe to another type of pipe is to use threaded adapters. Cut the CPVC, usually with a hacksaw, and install a tee. Then install a short piece of CPVC pipe and a threaded adapter into the branch outlet of the tee. Once the threaded adapter is in place, use another threaded adapter to mate any approved material with the CPVC pipe. If you will be converting from CPVC pipe to copper pipe, avoid soldering close to the CPVC. Solder a threaded adapter onto a length of copper and allow it to cool. When the soldered joint is cool, apply pipe dope and screw the two threaded adapters together. If you screw the copper adapter onto the CPVC before soldering in a length of pipe, the heat from the torch will melt the CPVC material.

Insert adapters

When you work with flexible plastic pipes, use insert adapters to convert to different pipe types. Cut the existing pipe, usually with a hacksaw, and install an insert type tee that is held in place by clamps or crimp rings (Fig. 15-8). The branch outlet of the tee may be an insert fitting or a threaded fitting. These combinations allow you to adapt to all types of pipes.

Fig. 15-8. Insert tee in polyethylene pipe.

ABS and PVC adaptations

When you wish to combine PVC and ABS pipe, use a special solvent or some type of adapter. The adapters may be threaded adapters or rubber couplings. If you want to make solvent weld joints, there is a special cement made just for this purpose. Your local plumbing supplier should have the combination cement. If he doesn't, you can always use rubber couplings.

Mixing and matching

Combining new plumbing, even when the pipe material is different, to existing plumbing is not difficult when you have the proper tools and adapters. As you have read, threaded adapters and heavy rubber couplings allow you to mix and match almost any combination of materials. When you make connections between different types of materials, always use an approved adapter. When the work is inspected by the plumbing inspector, he will check to see that the proper connections have been made. Using the proper adapters will reduce the risk of leaks and future problems.

16

Removing
existing plumbing
and fixtures

THE REMOVAL OF existing plumbing systems and fixtures can be both frustrating and dangerous. For the inexperienced person, the removal process can be the beginning of the end of his plumbing endeavors. A large number of homeowners plan to do their own plumbing until they encounter the problems associated with removing the existing plumbing.

In this chapter, I am going to share some of my professional secrets with you. Through many years of field experience, I have learned a number of techniques that simplify the removal of existing plumbing. It is not that these are dark, deep secrets, but many people would not think of using these methods unless they were involved routinely with plumbing. By sharing my experience, you will be a more efficient plumber. Let's take a look at how a master plumber can help you help yourself.

Toilets

The removal of a toilet can be simple, or extremely aggravating. When you look at a toilet, it appears that it would be fairly easy to remove. If everything goes in a text-book manner, it is easy to remove a toilet. Not all cases are text-book however, so the following paragraphs will help prepare you for almost any problem you may encounter.

Forgetting to cut off the water

Most people assume they will have to cut off the water before the toilet is removed. While this may seem common knowledge, not everyone will remember to cut off the water. If you attempt to remove the toilet without cutting off the water, water will flood the bathroom. This problem will be compounded if you find, while the bathroom is flooding, that the cut-off valve will not work.

Before you begin the removal process, cut off the water and test the cut-off valve to be sure it is working properly. Test the cut-off valve by flushing the toilet. If the tank does not refill with water, the valve is doing its job. If water leaks past the valve, cut off the main valve to the entire water distribution system. Something as simple as not cutting off the water may get your plumbing project off to a horrible start.

Uncooperative tank-to-bowl bolts

If you plan to remove the tank from the toilet bowl, the tank-to-bowl bolts must be removed. It is not unusual for the tank-to-bowl bolts to be deteriorated to the point where you cannot turn them with a screwdriver. If you are doing a complete demolition job, you could break the tank with a hammer. If you do this, wear safety glasses because pieces of china will go flying through the air. If you don't want to break the tank, you may have to cut the tank-to-bowl bolts with a hacksaw.

Before breaking out the hacksaw, try holding downward pressure on the head of the bolts with a screwdriver. Don't try to turn the bolt with the screwdriver, just maintain pressure on the bolt. With an adjustable wrench, attempt to loosen the nuts where the bolts penetrate the toilet bowl. Sometimes the pressure from the screwdriver will allow you to turn the nuts. If the nuts won't turn, you can try applying penetrating oil to the nuts. Let the oil set for awhile and try the procedure again.

If the nuts still will not turn, you will need a hacksaw. It may be difficult to get the hacksaw to work well in the tight space surrounding the nuts. A jab saw or even just a hacksaw blade may work better. If you use a plain hacksaw blade, wear gloves or wrap one end of the blade with duct tape (Fig. 16-1). The rough teeth of the blade can cut your hands during the sawing motion.

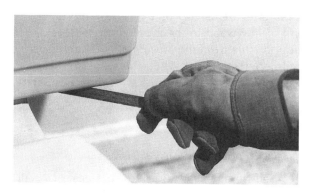

Fig. 16-1. Gloved hand using hacksaw blade to cut tank-to-bowl bolts.

Spinning closet bolts

The closet bolts are another possible source of frustration when you remove an old toilet. It is not uncommon for the bolt and nut to both turn as you try to loosen the nut. In many cases, the bolt will have been cut to a point flush with the top of the nut. When this is the case, there is nothing to hold when you try to turn the nut. If there is not enough bolt sticking up above the nut for you to hold, resort to more resourceful tactics.

Take a screwdriver with a wide blade and wedge the blade between the toilet and the metal washer that surrounds the closet bolt. If there is no washer, wedge the blade against the nut on the closet bolt. In either case, use the screwdriver as a lever and pry up against the nut to create tension. When you have tension, try turning the nut with an adjustable wrench. Many times this is all that is needed to loosen the nut.

When you simply cannot loosen the nut, you will have to cut the closet bolt. Use a hacksaw, jab saw, or hacksaw blade to cut the closet bolt. Hold the nut with a pair of pliers and cut the bolt off below the nut. If you are going to junk the toilet anyway, you can always break the base of the toilet for a quick removal. Remember to wear safety glasses if you break the toilet.

Residual water

Since toilets have built-in traps, they retain water in the bowl. When you lift the toilet bowl off its flange, the residual water will run out onto the floor. If getting the floor wet will cause a problem, use a plunger to force all the water out of the toilet before you remove it from the flange. A few good thrusts with a plunger is all it will take to force most of the residual water down the drain.

The tank on the toilet will also be holding residual water. Residual water will be built below the level of the flush valve. When you remove the ballcock nut, the water will rush out onto the floor. Use paper towels or a sponge to remove the water before it leaks onto the floor.

Lavatories

Lavatories are usually pretty easy to remove. The first step, for all lavatories, is to cut off the water to the faucet supplies. Next, use a basin wrench to loosen the supply nuts from the faucet inlets. Then disconnect the trap from the lavatory's tailpiece (Fig. 16-2). When the waste and water connections are disconnected, you are ready to remove the lavatory bowl. Safety glasses are an asset when you are working under the sink. They protect you from falling rust and other foreign objects.

Self-rimming lavatories

Self-rimming lavatories are the easiest of all lavatories to remove. Cut the

Fig. 16-2. Pliers (t&g) being used to loosen slip nut on a P-trap.

caulking around the rim of the lavatory with a knife. When the caulked seal is broken, push up from below the bowl to remove the lavatory from the countertop. That is all there is to the removal of a self-rimming lavatory.

Rimmed lavatories

Rimmed lavatories may be a little more difficult to remove. The clips that hold the bowl to the ring can be rusted or corroded to the point where they are hard to turn. You may use a penetrating oil to loosen the clips, but this takes time and may not work. Since you will not be using the ring or the bowl again, rip the clamp out with a pair of pliers. This method is fast and efficient. Once the clips are removed, you can drop the lavatory into the cabinet below the countertop.

Remember, the bowl will fall on you if you are under it when enough of the clips are removed. When possible, have a helper hold the sink from above as you remove the clips in order to avoid a headache. The helper can hold the weight of the sink by grasping the spout of the lavatory faucet.

Wall-hung lavatories

Wall-hung lavatories are simple to remove. Disconnect the waste and water connections. If the lavatory is equipped with legs, don't bother to remove them. When the waste and water lines are free, inspect the underside of the lavatory for lag bolts that may be holding the bowl to the wall. If you find these bolts, remove them, but most lavatories will not have them. Place a hand under the rim on each side of the lavatory and lift upward. The bowl should lift off its bracket without much trouble.

If the lavatory is stubborn, wiggle it from side to side and lift again. If it still won't cooperate, double check for lag bolts going into the wall. If there are no bolts securing the bowl, apply strong pressure from below the bowl. You can use your knee for this if you are working alone. If you have a helper, have the helper push up from below while you pull up from above. Normally, the removal of wall-hung lavatories will not cause you much grief.

Bathtubs

Removing bathtubs can be hard work. If you are dealing with a cast-iron bathtub, you may have to move over 400 pounds of bathtub. With one-piece fiberglass tubs, it may not be possible to get the old tub out of the house in one piece. Negotiating stairs with either of these tubs can be quite challenging. In all the following examples, you should turn the water off to the tub valves before you remove the tub.

Typically, the tub valve will be replaced with the tub, but even if you don't anticipate fooling with the tub valve, cut off the water. When you are wrestling with a tub, it can get away from you. In the process it may shear the tub valve off and flood the home.

Cast-iron bathtubs

Before you attempt to remove the tub, cut off the water and disconnect the tub waste and overflow. Before you remove the tub, assess the removal route and access limitations. Doors and stairways are two common hindrances when it comes to taking a tub out of the house. Since cast-iron tubs commonly weigh in excess of 400 pounds, most plumbers break them into small pieces with a sledge-hammer. Safety glasses are a must for this type of removal.

An 8-pound sledgehammer will destroy a 400-pound tub in short order. With most jobs, breaking the tub into pieces is the easiest way to remove it. If you wish to salvage the tub, you are in for a rigorous job. You will need at least one extra set of hands and a few more won't hurt. To remove the tub you will have to destroy the walls surrounding the tub to a point at least 2 feet above the tub.

Remove the walls you have cut. In most cases, you will have to remove the tub spout and the tub valve before the tub can be lifted out in one piece. When all the plumbing has been disconnected from the tub, you are ready for the removal process to begin. If you have a weak back, don't even consider doing this job yourself. There is the potential for a strained back, crushed fingers, and other bodily damage.

When the wall sections are removed, you should be able to see that the tub is sitting against stud walls. Reach into the back wall to get a grip underneath the tub's rim. Keep your hands in the middle of the wall cavity. If your hands are too close to the studs, and the tub slips, you might crush your fingers. Grip the tub rim and pull it up towards you. Don't allow it to fall forward onto your toes! The crushing effect of a 400-pound tub on your toes will stick with you for a long time.

When the tub is turned up and is resting on its apron, someone needs to crawl over the bathing unit. Ideally, you should have two people on each side of the tub. Since we don't live in a perfect world, I will assume you only have one helper. If the wall space will allow it, shift and walk the tub out of the recessed area. When the space is tight, and it usually is, the job is more difficult.

Arm your helper with a 4-foot section of a 2-×-4 stud to use as a lever. Rock the tub up so that the helper is able to place the stud under the tub. Use the lever to force the tub out, little by little, first one side and then the other. This will take time, but it will work, and it will reduce the risk of personal injury.

If you are impatient, you can try a bolder approach. Instead of using the lever and working the tub out gradually, you can roll it out. Turning the tub over and over will get it out, but you may cause several types of damage in the process. If you are working on a second floor, the weight of the tub falling on the sub-floor may damage the ceiling below. If your feet find their way under the tub, you will probably need medical attention. The slow way may take longer, but it gets the job done and with less risk all the way around.

Once you have the cast-iron monster out of its lair, you must get it out of the house. You will be able to get the tub through any standard door, hallway, and stairway, but remember its weight. Two people can carry a cast-iron tub, but it's no picnic. If you must walk the tub up or down steps, the risk for injury increases. Use common sense and don't place yourself in jeopardy during the removal of cast-iron tubs.

Steel bathtubs

Steel bathtubs are relatively light in weight and are removed using much the same methods described for removing cast-iron tubs. However, you cannot break a steel tub into pieces with a sledgehammer. Before you can tip a steel tub up off the stud wall, you may have to remove some nails from the nailing flange. It is not unusual to find roofing nails in the nailing flange.

One-piece bathtub and shower combinations

One-piece bathing units are usually made from fiberglass. The fiberglass construction makes these units light, but their size complicates the removal process. One-piece units are generally installed during the construction process. They are not meant to go through established living space. These large units will not fit through most doors and stairways.

To remove a one-piece, tub-shower combination, remove the tub spout and faucets. Next, remove the walls covering the nailing flange of the unit. When the nailing flange is exposed, remove the nails that hold the unit to the stud walls. Then grasp the unit by the rim of the tub and pull it out away from the wall. To get the massive unit out of the house, you will have to cut it into pieces. A reciprocating saw is an excellent tool for cutting these fiberglass units. A hacksaw will get the job done if you don't have a reciprocating saw.

Showers

Showers may be made from a host of materials. There are metal showers, fiberglass showers, tiled showers, and others. The removal of a shower is usually a bit easier than the removal of a bathtub. The following paragraphs will help you understand how to cope with the removal of various types of showers.

One-piece fiberglass showers

One-piece fiberglass showers will be removed in much the same way as a one-piece tub-shower combination. After the water has been cut off, remove the shower valve and disconnect the shower drain from the trap. To disconnect the drain you may have to cut into a ceiling below the shower drain. When the plumbing is disconnected, follow the same removal procedures as described for one-piece fiberglass tub-shower combinations. Since some showers are small, you might get the unit out of the house without cutting it into pieces. With most jobs, you will be forced to butcher the shower in order to remove it from the bathroom.

Metal showers

Metal showers are removed a little differently than fiberglass units. As described above, disconnect the waste and water lines. The metal shower will normally be held together by nuts and bolts. Disassemble these nuts and bolts to reduce the shower to a manageable size. These take-down units are easy to remove.

Tile showers

Tile showers require a totally different removal procedure. The walls of the shower will be made of tile that generally is attached to drywall. Wear safety glasses when you work with these units. Tile is brittle and if broken, the tile may fly into your face. As described above, disconnect all plumbing.

The tile can be removed in several ways. You can smash it with a hammer. You can pry it off with a wood chisel and hammer, or you can cut the drywall and take it down in sections. Before you can remove the shower base, you must remove the tile walls. Remove the tile and you will be able to work with the base.

The shower base may be made of concrete, fiberglass, or molded stone. If the base is tiled, you can expect to find concrete below the tile. If the base is stone molded or fiberglass, you will find a nailing flange when the wall tile is removed. After the wall coverings are gone, if you have a molded base, simply lift it out of the recessed area. If you have a tiled, concrete base, you will have to break it up. This can be done with a sledgehammer. Again, don't forget to wear safety glasses.

To break out a concrete base, hit it with a sledgehammer until you win the battle. If there is living space below the shower location, be aware of the effect the pounding will have on the ceiling and light fixtures. It is embarrassing to have a chandelier, worth hundreds of dollars, destroyed during the removal of a shower base. WIth a big hammer and a little sweat, getting the concrete base out will not be a problem.

Kitchen sinks

Kitchen sinks rarely present major removal problems. Sometimes the sink clips will be seized, but you can twist them out with a good pair of pliers. Once all the waste and water fittings are disconnected, removing the sink should not be a problem. If you are removing a cast-iron sink, you will enjoy the aid of an extra set of hands. A double-bowl, cast-iron sink is heavy. With the help of a friend, the task will be much less exhausting.

Bar sinks

Bar sinks are removed with the same techniques used on kitchen sinks. Disconnect the waste and water connections. Then remove any sink clips and lift the sink out of the counter. These small sinks rarely produce problems for the plumber.

Laundry sinks

The laundry sinks you may run across might be made of various materials. The laundry tub may hang on the wall or be supported by legs. In either case, unless you are working with a concrete laundry tub, removal procedures are simple.

Plastic and fiberglass laundry tubs

Disconnect the waste and water pipes. As in all of these examples, be sure the water is cut off before you disconnect the water connections. For laundry tubs supported on legs, removal will be as simple as picking up the basin and moving it. Plastic and fiberglass laundry sinks are not heavy and may be relocated by a single individual of average strength.

If the laundry tub is hanging on the wall, it will be attached to a wall bracket, like a wall-hung lavatory. To remove this style, just lift it up off the bracket and set it aside. Fixture removal does not get much easier than with these lightweight laundry tubs. Some laundry sinks have faucets that sit on the back-splash rim. If your faucets meet this description, they may have to be removed prior to moving the sink. In some cases, you can slide the sink out from under the faucets. In others, the faucets must be removed.

Concrete laundry sinks

Concrete laundry sinks are looming grey monsters with enough weight to test anyone's strength. It is senseless to attempt to move these mammoth fixtures alone. If you will be eliminating a concrete laundry tub, you will have to break it into pieces or have strong help.

These dinosaur fixtures are unlike any other you are likely to encounter. Typically, a metal frame will hold the concrete laundry tub. The unit may have a single or a double bowl. The first step in the removal process is to disconnect all the waste and water lines. These old-style tubs almost certainly will have faucets that mount on the back-splash rim. The odds are high that you will have to remove the faucets if you hope to take the sink out in one piece.

There are times when the metal frame will have rusted and become weakened. If the frame is weak, it is possible that these heavy sinks will fall when disturbed. Be careful. If these sinks land on your feet, you will surely wind up in a cast. Most plumbers remove these beasts in pieces. They do this by breaking them into manageable pieces with a sledgehammer. Safety glasses are a must when using a hammer to demolish a concrete laundry tub.

There have been few occasions in my career when I have removed this type of sink in one piece. On those occasions, after the job was done, I wondered why we didn't just break the unit into bits. If you plan to haul the big sink out in one piece, make sure you have strong, dependable help. If anyone slips or gives in to the weight, someone is likely to get hurt. I can think of no reason to salvage a concrete laundry tub. Unless you have a determined reason, break the sink into pieces you can handle with ease.

Faucets

Faucets are one of the most often replaced fixtures and they can give you some of the worst times imaginable. The faucets for tubs and showers are usually easy to replace but sink faucets can give you a rough time. When you are ready to remove a faucet, cut off the water supply. Since you will normally be on your back and looking up, wear safety glasses. Rust and other debris will fall from above and get into your eyes if you are not careful.

Kitchen faucets

Kitchen faucets are probably the worst of the bunch to remove. It seems these faucets are always rusted and hidden behind a maze of drainage piping. Faucet replacement requires you to remove the supply nut, where the supply tube is connected to the faucet inlets. With a basin wrench, this step is usually not too demanding. It is the mounting nuts that can drive you crazy.

The mounting nuts are the ones that hold the faucet onto the sink. There is ordinarily a ridged washer between the nut and the sink. These nuts are famous for rusting and becoming all but impossible to turn. When this happens, you are at a disadvantage. Due to their location, mounting nuts are hard to see, reach, and turn. Without a basin wrench, you have little chance of loosening a seized mounting nut (Fig. 16-3).

Fig. 16-3. Adjustable basin wrench on the supply nut of a kitchen faucet.

Even with a good basin wrench, there will be times when the nuts just will not turn. When this is the case, you must cut the nuts off the faucet inlets. If you have easy access to the nuts, this job will not be so arduous, but you will not have easy access. The nuts will be hidden behind the sink's bowls and drainage fittings. To say the least, the working conditions will be poor.

You have two viable options under these conditions. You can saw off the nuts or cut them with a hammer and a chisel. Fortunately, when the nuts are heavily rusted they are usually weak. A few direct hits with a hammer and chisel should pop them right off the threaded portion of the faucet. When you use a chisel, be careful not to harm the sink. The impact applied to the nuts is enough to damage inexpensive, stainless steel sinks. If you decide to use a hammer and chisel, do so with care.

If you have a reciprocating saw, and enough room, use it. Install a metal-cutting blade and saw through the nuts. Many inexperienced plumbers try to saw off the threaded portion of the faucet. It is difficult to get good saw placement to cut the faucet shanks and there is much more metal to cut in the faucet than there is in the nut. When all you have is a hacksaw, you will have to work harder, but you can get the job done. With most sinks, you will not have room to work the hacksaw. You will have to remove the blade and use it without the saw's frame. If you have a jab saw that accepts a metal-cutting blade, it will work well under these conditions. Once you cut through the nuts they should fall off and allow you to remove the faucet.

Other faucets

Lavatory faucets are removed in the same way as kitchen faucets. Lavatory faucets aren't generally as hard to remove as kitchen faucets. The faucets for bar sinks and many laundry tubs will be removed with the same methods. For laundry sinks with rim-mounted faucets, cut the supply pipes before you remove the faucets. In better installations, there will be unions in the supply pipes that will allow an easier faucet removal.

Water heaters

Removing a water heater can become an interesting proposition. On average, these water tanks customarily hold between 40 and 52 gallons of water, depending on the size of the water heater. If you cannot get the tank to drain, trying to move it will be quite a chore. As water heaters build up sediment, their drains sometimes block. This condition makes draining the tank with familiar methods slow or impossible. When you are dealing with a cantankerous water heater, you must get creative.

The first step when removing a water heater is to cut off the electrical power or fuel source. Second, cut off the water line feeding the heater. Have the electrical wires disconnected by a licensed electrician, or the fuel source disconnected by the appropriate licensed technician, before you proceed with the plumbing. Cut both the hot and cold water pipes about 6 inches above the water heater. Connect a hose to the heater's drain and then open the drain by turning it counterclockwise. As the water begins to drain, open the relief valve. Cutting the water pipes and opening the relief valve allows air to enter the tank. This will facilitate a faster draining.

If the drain appears plugged, try to rod it out with a piece of wire; a coat hanger will do fine. If the rod doesn't make a significant difference, you will have to take more drastic measures. In these next steps, the area surrounding the water heater will get wet. If flooding the area is prohibited, don't attempt to speed up the slow draining.

Put a pipe wrench on the pipe nipple to which the drain handle is attached. Turn the nipple counterclockwise. After a few turns, the nipple will come out. This should allow water to rush out of the water heater. If the water still will not drain, probe the opening with a screwdriver. Only in rare instances will the sediment block the opening to a point where this technique will not clear the opening. If for some reason this procedure doesn't work, don't despair. You have another option.

Remove the cover plates that conceal the water heater's elements. Be very sure the electrical power to the water heater has been shut off before doing this. There is high voltage behind these panels when the power is on. Start with the lower element. If the element is screwed into the tank, you will need an element wrench. In a pinch, you can turn these screw-type elements with a basin wrench. If you have bolt-on elements, remove the bolts.

Using whatever tool you can, remove the element. I have never seen a water heater so filled with sediment that I couldn't get water to flow from the element hole. As the tank empties you will have to lay it on its side to drain the water below the element hole. This may not be the easiest way to drain a water heater, but sometimes it is the only way.

If you cannot use these methods because of the flooding, you will need a pump. Insert a hose or tube into the water heater and pump the water out. The insertion may be made from one of the water connection locations or at the relief valve opening. Water heaters can be tough, but with these methods you will win the battle.

Drains

Removing old drain pipes may require some thought. Since most of the pipes you will be removing will be made of cast iron or galvanized steel, they will be heavy. Without the proper tools, these materials are hard to cut. There will be times when the drains are installed between the floor or ceiling joists. Trying to remove rigid pipe from the joists may prove fruitless. The first step in the efficient removal of drain pipes is the procurement of the proper tools.

Tools

When you cut out galvanized drains, a reciprocating saw with a metal-cutting blade is the best tool for the job. Keep several high-quality blades on hand. When you cut metal pipes, broken blades will be a common experience. The steel pipe has a tendency to pinch and break the blades as you cut. Inspect the saw and its electrical cord thoroughly. You may find water in the pipes as you are cutting. Without a well insulated saw, you could receive a bad electrical shock.

For cast-iron pipe, the best cutting tool is a rachet-style soil pipe cutter. These cutters will do an excellent job in areas with limited space. Without this type of cutter, you will be in for a long and tiring job. If you are doing complete demolition work, you may use a sledgehammer to break the pipe. Anything less than these two methods will be frustrating and very time consuming.

Leaving the old pipe in place

Sometimes you will be able to leave the old drains in place. If you can, you may save yourself several hours of hard work. Assess your options for installing new piping, and you may find you can work around the old piping. There are many circumstances that will allow you to install new drains without removing the old ones.

Cutting the pipes

Before you cut the old drain pipes, look to see what will happen when they are cut. Be careful not to put yourself in a position where the old piping will fall on you after it is cut. If you are removing a vertical stack, check to see if the weight of the stack will come crashing down around you when it is cut. Remember that there may be water standing in the pipes. Take adequate precautions to prevent electrical shocks. Verify the cutting area all around the pipe. Many times electrical wires will be installed close to pipes. If you fail to see these wires, and cut them with your saw, sparks will fly.

Lead closet bends

In the old days, closet bends were frequently formed from soft lead. It was not unusual for a lavatory or a bathtub drain to be piped into the side of these lead closet bends. If you are remodeling and find this type of arrangement, you should replace it. Lead bends usually are discovered when you replace a toilet or attempt to connect to the existing pipe that accepts the drainage from the toilet.

Lead closet bends are typically caulked into a cast-iron hub with a lead joint. The lead may have a side inlet that allows a smaller drain to enter into the lead bend. The lead will be shaped like a quarter bend and flare out at the base of the toilet. The flared lead acts as a type of closet flange. As a lead fitting ages it develops leaks. You should not attempt to connect a new installation to a lead closet bend.

The removal of the lead closet bend is not difficult. The lead is soft and may be cut with a hacksaw. The lead is easy to cut, but is heavy. When you cut the fitting, its weight will allow it to bend at the cut and swing into your body. This may be uncomfortable under any condition, and it may be especially dangerous if you are working on a ladder. Support the lead to prevent it from hitting your body.

Once the lead is cut, chisel it out of the cast-iron hub. Sometimes it is easier to cut the hub off the cast-iron pipe and convert to plastic pipe with a rubber coupling. If there is insufficient pipe for this, you will have to clear the hub of all the existing lead and oakum. When the hub is empty, you may use a rubber adapter ring to convert to plastic pipe.

Lead traps

Lead traps were used for many years and are still found on some bathtubs and showers. If you are replacing a fixture that has a lead trap, replace the trap and the trap arm. You may be able to cut the lead trap arm close to the cast-iron hub and use a rubber coupling to extend the replacement piping. If not, use the same methods described for lead closet bends in order to clear the cast-iron hub and make the new connection.

Vents

The removal of vents will go about the same as drains. Be especially aware of what will happen when the vent pipes are cut. If these pipes are not supported, they may fall and injure you or damage the home. Don't forget to repair the roof after the vent pipe is removed. I have seen many ceilings damaged by water from old vent holes that were never sealed.

Water pipes

Water pipes normally are easy to remove because they are relatively small. There are many times when the removal of existing water pipes will not be necessary. You can often run new pipes without removing the old ones. If you use an electric saw to cut the pipes, be sure it is well insulated against electrical shocks. It should go without saying, but remember to cut off the water to the system before you remove the water pipes.

Summary

Removing existing plumbing is frequently a part of any remodeling job. When you are equipped with the proper tools, this work should go smoothly. Use common sense in your approach to the task. Evaluate the circumstances, look to see what will happen with each step you take, and proceed with caution. Any time you are working with old materials, you may be taken by surprise. While you are turning on a pipe at one end of the basement, a pipe at the other end may rupture. Good planning is the key to a successful removal of old plumbing without regrettable results.

17

Remodeling installation variations

WHEN YOU ARE WORKING with existing plumbing, you are likely to experience some unexpected results. Jobs never seem to go as planned when they involve old plumbing. Jobs that are expected to take less than an hour of your time may wind up requiring several hours of effort. You can eliminate some of these inconveniences with good planning and careful inspections.

Offset closet flanges

When you replace an old toilet with a new one, the existing drain may not be in the right place. If the drain needs to be moved several inches, you will have to relocate the drainage piping. If it only needs to be moved an inch or so, you can use an offset closet flange (Fig. 17-1). This inexpensive fitting may allow you to avoid relocating the drainage piping.

Offset flanges install like any other closet flange. The difference is in the shape of the fitting. Instead of having the drainage hub drop straight down from the flange, it is offset. The offset allows you to move the bolt location for the toilet to either side, or from front to back. When you have a floor joist in the way, or an existing drain you don't want to relocate, offset flanges may be the answer to your dilemma.

Pipe chases

If you have explored every avenue and cannot find a way to run the pipes in the existing walls, consider building a pipe chase. Find an unobtrusive area and build a box around the new piping. Closets frequently offer the possibility for unnoticed pipe chases. If there are no closets to use, get creative. If you are forced to build a chase in the open area of a room, give it an additional purpose for being there. You could recess shelves into the chase to act as a bookcase. You

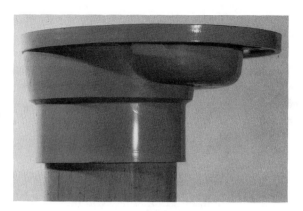

Fig. 17-1. Offset closet flange.

might design the chase to act as an entertainment center that will hold your stereo, television, and VCR. With a little thought, a chase will make your plumbing job easier and may add to your home's usefulness.

Plumbing capacity

One of the biggest concerns when adding new plumbing is the capability of the existing sewer and water source to provide adequate service. Before you spend thousands of dollars on a new bathroom, make sure it will function properly with the existing conditions.

Adding a basement bathroom

Putting a bathroom in the basement might cause a problem when it comes time to vent it. The bathroom must have a vent that goes up into the living space above the basement. You may take the vent either directly through the roof or tie it into another suitable vent. If you are tying into an existing vent, you must tie in at least 6 inches above the flood-level rim of the fixture served by the existing vent.

Most people never consider the need to vent their new basement bath. Except in rare circumstances, some wall in the main part of the home will have to be opened for the vent. An experienced plumber may be able to get a vent up without cutting a wall, but the odds are against it. When you plan the basement bath, plan the routing of the vents.

Size limitations on bathing units

When adding a bathroom within an existing house, you must be aware of access limitations. The size of the ingress areas will dictate the maximum size of the fixtures. For example, one-piece, fiberglass tub-shower combinations will not fit through the doors and stairways of most homes. One-piece units are used almost exclusively in new construction and add-on additions. Under new-construction conditions, these bathing units can be set in place before walls and doors are installed.

Even if you are able to get one of these units in the front door, it probably will not go around turns or up stairways. To compensate for this problem, you will have to look to sectional bathing units. There are many types of sectional units available. There are two-piece, three-piece, four-piece and five-piece units on the market. With so many to choose from, you should not have any difficulty finding one to suit your needs.

Modernizing tub and shower faucets

If you are doing light remodeling, you may not plan to replace the bathing unit. While the bathing unit may be spared, the faucets will frequently be replaced. How will you replace a two-handle faucet with a single-handle faucet? The answer is simple—you will use a remodeling escutcheon. These special escutcheons are designed to allow the installation of a single-handle faucet where previously there was a multi-handle faucet. Once the old faucet is removed, the new, single-handle body is installed. The remodeling plate fits over the body and is large enough to cover the holes left from the old faucet.

Garbage disposers and septic fields

When modernizing a kitchen, a garbage disposer is often planned for the new kitchen. We have already talked about how old, galvanized drains may cause problems when a disposer is added. Now, we look at a question that is difficult to answer. Is it wise to install a garbage disposer in a home with a septic system? Depending upon who you ask, you will receive conflicting comments on this question.

Some jurisdictions prohibit the installation of disposers in homes with septic systems. Other locations allow the use of disposers with septic fields. There are arguments that go both ways. What is the truth of the matter? Well, I don't know if there is a cut-and-dried answer. I have seen many homes successfully use disposers with septic systems. I cannot recall a time in my experience when a disposer was directly related to a septic system failure.

The argument for not combining garbage disposers and septic systems is one of logic. It is said that disposers add to the responsibilities of a septic system. Further, the food particles from disposers are accused of not breaking up and dispersing well through the septic field. There are opinions that say the waste from a disposer will clog the septic field, and result in expensive repairs or the full replacement of the field. This opinion makes sense when you think of what goes through some disposers.

In my opinion, much of the final outcome is predicated on the type and amount of use given the disposer. If the appliance is used, but not abused, I doubt you will have a problem with the septic field. If you allow grease and other objectionable substances to make their way to the septic tank, you are asking for trouble. Plumbers affectionately refer to garbage disposers as pigs, but they are an appliance, not a farm animal. Be selective in what you feed the pig, and chances are you will not have significant problems.

The keys to success

The keys to success when you install plumbing for remodeling jobs are planning, anticipation, and flexibility. As a remodeling plumber, all of these responsibilities rest on your shoulders. You must evaluate the existing system, anticipate problems, design a working plan, and be flexible in the plan's execution. Remodeling jobs rarely go the way they are initially planned. You must be willing, and able, to roll with the punches.

If you enter into a remodeling job under the false assumption that it will be simple, you will regret ever tackling the job. When you devote enough time to do the job right, you will eliminate most of the problems.

Part IV
Troubleshooting pointers

18

Common malfunctions with DWV systems

EVEN AFTER THE JOB is done and has been approved at the final inspection, you might experience problems. There are numerous complications that may arise after the job is finished. The drain-waste-vent (DWV) system plays an important role in the use of the plumbing system. When the drains and vents in the job fail to work, the rest of the plumbing is paralyzed. If you installed the plumbing, it will be your responsibility to correct the deficiencies in the system. This chapter will provide you with help in locating and correcting system failures.

Vent problems

Most vent problems appear as a drainage problem. With the exception of escaping sewer gas, vent failure will affect the drainage system. When a vent is obstructed, drains will not drain as quickly as they should. This is the most common fault found in the vent system. Vents provide air to the drains, which enables the drains to flow freely. When vents become clogged, the drains don't work as well as they should. If there is a break in the vent piping, sewer gas may enter the home. The unpleasant odor associated with sewer gas will be the best indicator of a broken vent or a dry trap. Normally, these two potential failures are the only ones encountered with the venting system.

Sewer gas

Sewer gas is potentially dangerous. It may be harmful to your health when inhaled, and it is explosive when contained in un-vented conditions. Sewer gas is essentially methane gas. The gas is formed in septic tanks and sewers. When the traps and vents in your home are working properly, you will not notice any sewer gas. If there is a defect in these systems, you may notice an unpleasant smell. Some people fail to identify the odor and develop health problems because of the escaping sewer gas. In small quantities, sewer gas offers minimal risk, but in large quantities, especially in confined conditions, the gas is potentially fatal. If you experience an indistinguishable odor in your home, investigate the plumbing system.

Sewer gas can invade the home from many sources. When the trap on the fixture goes dry, sewer gas can infiltrate the dwelling. If there are holes or bad connections in the vent piping, sewer gas can escape. If there is a vent terminating near a window, door, or ventilating opening, sewer gas may be pulled into the home through the opening. If the plumbing was installed in compliance with the plumbing code, there is little chance of the gas being sucked in from windows, doors, or ventilating openings.

If the vents were tested during the rough-in inspection, the joints in the pipe should be sound. However, there is always the possibility that a nail was driven into one of the pipes after the inspection. Nail plates are supposed to prevent this type of problem, but circumstances might prove differently. The most common cause for escaping sewer gas is dry traps. Since traps are a part of the drainage system, they will be covered later in this chapter.

Clogged vents

If the vents become obstructed, the drains will not drain well. When vents become clogged, leaves are a prime suspect. If leaves fall into the vent pipe, they might become lodged in fittings and build up. With enough leaves in it, the vent can become useless. Occasionally, a bird will build a nest on, or in, a vent pipe. The nest blocks the flow of air and disables the vent. These two situations do not occur often, but if the vents fail, check for these types of stoppages in the pipe.

Capped vents

When the installation of drains and vents is tested with air pressure, the vents are capped. The caps are installed on the end of the pipe, where it terminates through the roof. I have had occasions when my plumbers failed to remove the vent caps after testing the system for leaks. If the vents fail to operate correctly immediately following the installation, check to see if the pipes are capped.

Frozen vents

In cold climates, vent pipes sometimes freeze. The moisture accumulated in the pipe will begin to freeze on the walls of the pipe. If the temperature remains cold for an extended period, the ice will build up until it blocks the pipe. Cast-iron and galvanized pipes are more likely to freeze than plastic pipe. Unless the temperatures are brutal for a long time, it is unlikely the vents will freeze to an extent that hampers the operation of the plumbing system.

Drain complaints

After a new system is installed, the drains may present some interesting problems. Finding the flaws in a new system is not difficult when you know what to look for and how to look for it. Many of the problems are only an inconvenience, but some of them can shut the system down. By following the proper troubleshooting techniques, you can find and fix these faults.

Drains that drain slowly are one of the most frequent complaints with the DWV system. Slow-moving drains may be the result of a number of possible problems. The type of drain experiencing the problem has a lot to do with how you troubleshoot it. Let's look at some examples of how you may correct sluggish drains. Since vent problems have already been discussed, they are not dealt with in the rest of this chapter. However, if you are having drainage problems, don't neglect the possibility that the problem is with the vent, not the drain.

Bathtub drains

Bathtub drains may run slowly for several reasons. The waste and overflow might need to be adjusted. There might be an accumulation of hair built up in the tub waste. The trap could be obstructed by foreign objects. The drain's vent might be obstructed. All of these reasons, and a few others, can cause the bathtub to drain slowly.

Tub waste and overflows. If you are having trouble with a newly installed bathtub, the problem is most likely in the tub waste and overflow. There are several types of wastes and overflows. If you have the type where you flip a lever to hold and release water from the tub, it may need to be adjusted. The length of the rod on these units is adjustable for tubs with various heights. If the rod has been set at a length too long for the tub, the tub may drain slowly.

To determine if the tub waste needs to be adjusted, remove the cover from the overflow pipe. Remove the assembly with the rod and plunger on it. Next, remove the pop-up plug from the tub's drain, if it has one. Block the tub's drain with a rag and fill the tub with water. Remove the rag or stopper and see if the tub drains properly. If it does, you need to adjust the tub waste. If it does not, the problem is farther along in the drainage system.

Adjusting the tub waste is simple. The rod that you have removed will have threads and nuts on it. The housing it screws into will usually be marked with various numbers. These numbers represent the height of the tub. If you are adjusting the waste to allow the water to flow more freely, you will want to shorten the rod. To do this, loosen the set nuts and screw the rod farther into the housing. You may have to achieve the proper setting through a trial-and-error method. Just keep adjusting the waste rod until the tub drains satisfactorily. You have proven that the problem is in the tub waste. All you have to do is reach the proper setting and the tub drain will work fine.

If the tub waste does not have a trip lever, it most likely is controlled by a stopper plug. These plugs are usually depressed to open and close the tub drain.

At times, the rubber gasket on the stopper might impede the flow of the draining water. When this is the case, turn the stopper counterclockwise for a few turns. With continued adjustments, you should reach a point where the drain functions properly.

Bathtub traps. Bathtub traps are usually "P" traps. If you have determined that the problem is not associated with the tub waste and overflow, the trap is the next logical place to look. With new installations, there is rarely a problem with the trap, but there could be. Run a small spring snake down the overflow pipe of the tub and into the trap. If the snake meets no resistance, the trap should not be causing the problem.

The bathtub drainage pipes. If you are dealing with new plumbing, there is no reason why you should be having trouble with the bathtub's drain pipe. However, if you installed your new tub by connecting to an old piece of galvanized pipe, the drain pipe could very well be the problem. Continue to put the snake through the trap and into the drain pipe. For some blockages, a hand-snake will clear the pipe. If you have a strong blockage in a galvanized pipe, you will need an electric drain cleaner to eliminate the clog.

When galvanized pipes become blocked, the stoppage often consumes the entire pipe. Small drain snakes will only punch holes in the clog. The drain will work better for awhile, but then it will stop up again. To remove the typical stoppage from a galvanized pipe, you will need an electric drain cleaner with a cutting head. As the head rotates its way through the pipe, it cuts away the debris from the sides of the pipe. Unfortunately, even a thorough cutting of the blockage is often only a temporary repair. The rusted interior of galvanized pipe will soon catch new grease, hair, and other particles and create a new clog.

Shower drains

Showers do not have a waste and overflow arrangement to adjust. When shower drains move slowly, you know the problem is in the trap or the drain line. The same methods described for clearing blockages in tub traps and drains apply to showers.

Sink drains

If the sink is draining slowly, you can assume the problem is in the trap or the drainage piping. Unlike tub and shower traps, the traps for most sinks are readily accessible. These traps are frequently joined with nuts that allow you to disassemble the trap. When the sink drains slowly, disassemble the trap and inspect for any obstruction that may cause the slow draining of the sink. If the trap is clear, run a snake down the drain pipe.

It is unusual for a sink drain to operate poorly following a new installation. If the sink is a lavatory, you might have a problem with the pop-up assembly. The pop-up may need to be adjusted to allow the water to drain faster. It is also possible that you have a hair clog on the pop-up rod. To test these possibilities, you must remove the pop-up plug. If you have a hair clog, you will see it when you look down the drain assembly.

To determine if the pop-up needs adjusting, place a rag or removable stopper over the lavatory drain. Fill the lavatory bowl to capacity and remove the rag or stopper. If the water drains well under these conditions, the pop-up assembly needs to be adjusted. If the water continues to leave the bowl slowly, the problem is in the trap or drainage pipe.

Slip nuts. The drainage connections under a sink are often made with slip nuts and washers. It is not uncommon for these connections to develop small leaks in the first few weeks of operation. If you notice water dripping from the slip nut, tightening the nut should solve your problem.

Compression fittings. Compression fittings often need to be tightened after the first few days of use. Until the joints have settled in, they may leak from pipe vibrations or movement. When a compression fitting is leaking, tightening the compression nut will usually correct the leak. If the leak is a bad one, you may have to replace the compression sleeve or nut. Sometimes the nuts crack and cause water to spray about. In either event, compression leaks are not difficult to fix.

Basket strainers

Normally, the only problem you may have with the basket strainer in a kitchen sink is a poor seal where the drain meets the sink. Remove the drain and reinstall it with a good seal of putty to stop the leak. If the sink will not hold water, the seal on the basket may be defective. If you suspect the basket is bad, replace it with a new one.

Water mysteriously leaving the sink. If you fill the sink with water, only to have it slowly drain out, you may have a putty problem. When the pop-up assembly seems to be properly adjusted, but water still leaks out, there may not be a good seal around the drain. If you installed the drain without sufficient putty, water will slowly leak out of the bowl and into the drain. This type of leak may be hard to locate. Since the water leaks into the drain, it doesn't show up under the bowl. To remedy the problem, remove the drain assembly and reinstall it.

Toilet drains

Toilets with a lazy flush may be affected by a number of problems. The wax seal might not be seated properly. The flush valve might not be operating properly, or the flush holes might be clogged. If all of these inspections pass muster, the problem is in the drainage pipe.

Flush holes. When you have a toilet that is not working the way it should, you must determine the location of the problem. Even though the toilet is new, don't rule out the flush valve or the flush holes. The flush holes are located inside the toilet bowl, under the rim. When you flush the toilet, water should wash down the sides of the toilet bowl. If it doesn't, check the flush holes. The holes may be plugged with minerals and sediment. It is also possible that the holes were not formed properly when the toilet was made. To check the flush holes, you will need a stiff piece of wire, like a wire coat hanger.

You can locate the holes with a mirror or by running your hand around the underside of the toilet's rim. Push the wire into the holes. If the wire goes in easily, you can rule out the flush holes. If you hit resistance, work the wire back and forth. If white crystallized particles fall out, you have mineral buildup. If the wire will not go in at all, you may have a defective toilet bowl.

Flush valves. For the next step, remove the cover from the toilet tank. Inspect the water level in the tank. If the water is at the fill line, which is etched into the tank, you have adequate water for a normal flush. While you have the cover off the tank, flush the toilet. Watch the flush valve when the tank empties. The flapper or tank ball should stay up long enough to allow most of the water to leave the tank. If the flapper or tank ball closes the flush valve early, there will not be enough water for a satisfactory flush. When this is the case, adjust the interior parts of the toilet tank. Instructions for these adjustments are covered in Chapter 20.

Wax rings. If all aspects of the toilet pass inspection, you must look to the wax ring or the drain pipe. Wax rings that are not installed properly might cause the toilet to flush slowly. In new installations, wax rings may be the primary cause of lazy toilet flushes. The problem occurs when the toilet is set on the flange. If the toilet bowl is not positioned directly over the wax ring, the wax may be compressed and spread out over the drain pipe. Since you cannot see the wax seal when the toilet is set, the bad seal goes unnoticed.

As the toilet is used, toilet paper catches on the wax extending over the drain. After several flushes, the build-up can hinder the flushing action of the toilet. To confirm the condition of the wax ring, remove the toilet from its flange. As an alternative to pulling the toilet, you may use a closet auger to check the toilet's trap and drain (Fig. 18-1). If you run the auger through the toilet, you may retrieve wax on the end of the auger. If you do, you can bet the wax ring did not seat properly during the installation. There is more on the use of closet augers in Chapter 20.

While the closet auger may reveal evidence of a faulty wax seal, the only way to be certain of the seal is to remove the toilet. If the hole at the bottom of the toilet has wax spread over it, you have found the problem. Scrape the old wax off and reset the toilet using a new wax ring. Be diligent so that you achieve a good seal when you reset the toilet.

Toilet drain pipes. If you have completed all of these troubleshooting procedures and still have a slow-flushing toilet, the problem is in the drain pipe. If you installed all the piping in compliance with the plumbing code, there is no reason why a new system should be blocked. When you have a stoppage, use a

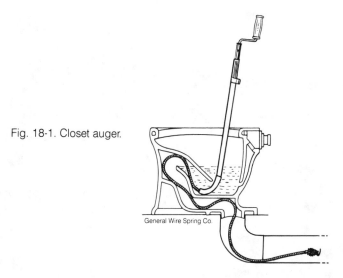

Fig. 18-1. Closet auger.

General Wire Spring Co.

snake to locate and clear the blockage. During construction and remodeling, some strange objects find their way into open drain pipes. This is especially true of the drains for water closets. Since these pipes are at floor level, it is easy for foreign objects to fall down the drain.

Start with a flat-tape snake (Fig. 18-2). Feed the snake into the drain until you come in contact with the blockage. For small drains, a spring-snake will negotiate the turns in the pipe better than a flat-tape snake. When you connect with the blockage, attempt to push it through the pipe with the snake. If the blockage will not clear, you will have to try an electric drain cleaner or cut the pipe.

Fig. 18-2. Flat-tape snake.

For drains with a diameter of 2 inches or less, you will want to use a small electric drain cleaner (Fig. 18-3). These units are hand-held and often resemble an electric drill with an attached drum housing. For drains with diameters of 3

Fig. 18-3. Hand-held power drain cleaner.

inches or more, you will want a larger machine (Fig. 18-4). These large machines are potentially dangerous. When a clog is encountered, the cable of these powerful machines can kink causing severe damage to your body parts. Before using an electric drain cleaner, observe proper safety precautions.

Electric drain cleaners may be rented from most rental cleaners. When you rent the machine, request full instructions on how to operate the machine safely. In general, wear gloves, don't wear loose or dangling clothes, and never allow excess cable to build up outside of the drain (Fig. 18-5). Slack cable can become very dangerous when you make contact with a solid blockage. Choose a flexible spring head for the drain cleaner (Fig. 18-6). These heads are easier to feed down the pipe and are less likely to cause uncontrollable torque than solid cutting heads.

Drain pipes

So far we have talked about specific types of drains. Now we will look at drain pipes in general. You already know that if you have connected new plumbing to old galvanized pipes, the galvanized drains may need to be replaced. As far as

Fig. 18-4. Large electric drain cleaner.

Ridge Tool Co.

Fig. 18-5. The proper method for using an electric drain cleaner.

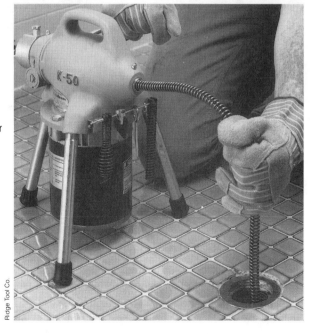

Ridge Tool Co.

Fig. 18-6. Spring-head attachment for electric drain cleaner. General Wire Spring Co.

the new pipes you have just installed, if they were installed according to code requirements they should drain properly. While you will probably not have any problems with the new drains, you will do well to know a few things to look for if you do.

Frozen drains

Drain pipes should not freeze because they are not supposed to hold water. In the drainage system, only the traps should hold water at all times. Even though drains shouldn't freeze, they sometimes do. If the drains are not working and resist the passage of a snake, you could have a frozen pipe. Obviously, this will only be the case in freezing weather.

It is difficult to diagnose a frozen drain, unless you can see the pipe. When drains are run in a crawlspace under a home, you have access to the pipes. Put your hands on the drains and you should be able to tell if they are frozen. When you cannot get to the pipes to touch them, you will have to make assumptions. If the pipe may be frozen, you can try a couple of different remedies.

If the pipe is draining, but doing so slowly, pouring very hot water down the pipe might melt the ice. If you have access to the pipe, a heat gun or a hand-held hair dryer might be sufficient to melt the ice. When the ice is far down the pipe, you may try filling the pipe with hot water and a garden hose. By attaching a garden hose to the hot water connection where the washing machine is, you can run hot water through the hose.

Keep an eye on the water hose. The hot water may cause the hose to overheat. Feed the hose into the drain that you suspect is frozen. Push it as far into the pipe as you can. When the hose reaches a stopping point, turn on the hot water. The hot water will force the water standing in the pipe out of the open end of the drain where the hose is inserted. As the hot water melts the ice, push the hose farther into the pipe. Continue this cycle until the drain is no longer clogged. If these methods don't work, you will have to call a professional. Professionals have steam machines they connect to the drain to thaw the ice.

Test balls and caps

Just as plumbers sometimes forget to remove the caps from vent pipes after a pressure test, they may also forget the caps on drainage pipes. Test balls may be left in a fitting in the crawl space or basement. It is possible that the thin plastic test cap was not removed before the trap was installed on a trap arm. A fitting will slide right over a plastic test cap, and these caps may blend in with PVC pipe and go unnoticed. If you have a total blockage of the drains, don't rule out forgotten test balls and caps.

Overloaded drains

When you have added new plumbing to an old system, it is possible the old drains are not adequate for the increased drainage load. For example, the existing sewer may work fine when one bathtub is draining, but may run slow when a second tub drains at the same time. The plumbing code is designed to prevent this type of problem. When you designed the new system, you should have computed the fixture load on the existing pipes. In doing so, the existing sewer should not be overloaded by the new plumbing.

If you are able to perform a visual inspection of all the drains, you can determine if they are large enough to handle a specified number of fixture units. The plumbing code assigns a unit rating to each fixture. Doing simple math will tell you if the pipe should work properly. Even when the numbers work, the drain may not. How could this be? Well, the drain may not be working to its full capacity for many reasons.

If there are roots growing into the sewer, the open diameter of the pipe is reduced. If the sewer is partially crushed, the flow on the pipe will be affected. If the sewer is not installed with the proper pitch, the pipe will not carry the waste as quickly as it should. All of these reasons might cause an old sewer to resist the demands of additional plumbing. These defects are hard to detect. You may have to expose all of the old piping to discover the cause for the slow-moving drains. In plumbing, you can never be sure of anything until you can see and verify all aspects of the system.

Full septic tanks

If the plumbing has been done in an existing home, the septic tank may be the cause of the backed-up drains. When septic systems are full, the drainage from the home has no place to go. Since the waste cannot enter the septic system, it remains in the pipe. When the pipes are full, the waste comes out of the drainage openings in the fixtures. If the drains are flooding back into the fixtures and a snake will pass through the pipes, the problem is most likely a full septic system. Under these conditions, call to have the septic tank pumped out.

Municipal failures

The failure of municipal sewers is rare, but it occasionally does occur. If none of the drains will drain, but a snake goes freely down the sewer, the problem may lie in the municipal sewer. Even though municipal sewers are very large pipes, they can still become clogged. If you cannot find a stoppage in any of the drains or sewer, call the public works department of the municipality. The public works department will send out a crew to check the main sewer.

The end of the line

Troubleshooting the drains and vents in the plumbing system is not difficult. If you follow a route of progressive system checks, you will identify the problem. Once the problem is known, you can correct it. Major problems with a newly installed DWV system are rare. When they occur, they can be found by trial and error. As you work your way along the DWV system, you will rule out possible causes for the trouble. In the end, you will find the defect in the system.

19

Routine
water pipe problems

AFTER A NEW WATER DISTRIBUTION system is installed, it may require some fine tuning. If new plumbing has been added to an existing system, the system may suffer from the alterations. Usually, there are not too many problems with the water distribution pipes. The pipes that do have problems, however, can cause big trouble. The potential problems range from disturbing noises to no water. This chapter will walk you through the troubleshooting methods used for water distribution systems.

No water pressure

One of the worst situations you might encounter with the water distribution system is when it will not produce any water. When you turn on a faucet and no water comes out, what do you do?

A lack of water might indicate a closed valve, a malfunctioning water pump, a broken water pipe, a frozen pipe, or some other mystery. Work your way along the system until you locate the problem. Troubleshooting the water pipes is a matter of logical progression. Start with the point where the problem arises and work your way through the system until the reason for the failure is found.

Closed valves

If you turn on a faucet and it will not produce water, check the cut-off valves that control the water supply to the faucet. If the faucet's cut-off valves are open and no water is coming out of the faucet, continue the search. Start by checking other faucets in the home. If none of the fixtures have water available, the problem will be found somewhere in the water main, water service, or at the water pump.

Electrical failures

Check all the valves in the main water line. If all of the valves are open, the problem is in the water service or the water pump. If you have a water pump, check the circuit breaker or fuse in the electrical panel that controls the pump. If the breaker is tripped or the fuse is blown, you have found the problem. If the electrical system is in working order, continue your quest.

Pressure tanks

Check the pressure gauge on the water storage tank (Fig. 19-1). If the gauge shows normal water pressure, open the boiler drain at the pressure tank. Sometimes pressure gauges stick and give false readings. When you open the boiler drain, you will be able to establish if you have water pressure at the tank. If you have pressure at the tank, you should investigate all of the cut-off valves again. When there is pressure at the tank, but none at the fixtures, the problem almost has to be a closed valve.

Fig. 19-1. Pressure gauge.

Bad pumps

If there is no pressure at the storage tank, you may have a defective water pump. Pumps may quit working at any time. To check out the pump, refer to Chapter 21. There could be a break or a blockage in the water service pipe between the house and its water source. If a new water service was installed, it may have been damaged when the trench was backfilled. It would be unusual, however, for the water service to be damaged to the point that no water comes through the pipe.

Municipal failures

If the water comes from a municipal source, the municipality may have cut off the water. They might be working on the pipes or have a break in the water main. After checking all of the valves and pipes, if you still have no water, call the municipal water department. They will check their system to see if their system is the cause of the problem.

Backflow preventers

Many jurisdictions require that you install backflow preventers on all water services (Fig. 19-2). Even if you are only remodeling, you may be required to install a backflow preventer in the main water line. Backflow preventers only allow water to flow through them in one direction. If the backflow preventer is installed backwards, you will not have any water pressure. If you have just installed a backflow preventer, check to see that it is installed properly. There should be an arrow on the side of the backflow preventer to indicate the flow of water. If the backflow preventer was installed backwards, you will have to turn it around to get water flowing to the fixtures.

Fig. 19-2. Backflow preventer.

Low water pressure

Low water pressure is a common complaint with water distribution systems. Finding the cause of low water pressure is more challenging than troubleshooting for no water pressure. Low water pressure might be caused by many different factors. The procedure for troubleshooting low water pressure is similar to the way that you check out a system that is not producing any water, but there are differences.

Municipal water systems

If the water comes from a municipal water system, the low water pressure might be a result of their piping. If they have a broken pipe, partially closed valve, or an unusually high demand for water, the water pressure may drop. Before you call the municipal water department, check out all of the possibilities of a failure with the system in the home. If, after a thorough check you don't find a cause for the low pressure, call the municipality and ask them to check their lines.

Private water systems

When the water comes from a well, the water pump and related equipment might be the source of the low water pressure. Chapter 21 concentrates on problems you might experience with water pumps and their associated equipment. Since this chapter is about water pipes, possible problems with pumps are not included here.

Valves

Partially closed valves might be responsible for low water pressure. When you are troubleshooting the system, check all the pertinent valves to be sure they are open and allowing a full flow of water. Turn the handle on the valve counterclockwise until it stops. When the handle is in this position, the valve should be open to its fullest extent.

Defective valves

It is possible that you have a defective valve. The valve may have been forged incorrectly or the washer may be defective. If the screw holding the washer on the stem of the valve works loose, the washer may obstruct the flow of the valve. This obstruction may cause low water pressure.

Solder in the valve

When valves are soldered onto water pipes, solder sometimes runs into the valve and creates a problem. If you are not experienced with soldering, you may put too much solder in the joints. If the valve is particularly hot, the solder may run into the valve and build up as it cools. The hump of solder might impede the flow of water. This type of problem is rare among professionals, but is not unheard of when soldering is done by inexperienced people.

Crimped supply tubes

The supply tubes that run from a stop valve to a faucet can be crimped easily. If the supply tube is severely crimped, it can reduce the water pressure available at the faucet. Normally, you will notice if you crimp a supply tube, but it is possible for the restriction to go unnoticed during the initial installation. When the water pressure is low at a single fixture, investigate the supply tubes.

Obstructed aerators

Aerators are the small objects that screw onto the spout of faucets. Aerators have a nylon diversion disk and screen wire in them. If the openings in these devices become plugged, the water pressure will be affected. When the aerator is clogged, water may spray out from the spout, or it may come out in a puny stream. This type of problem is faucet related and is covered in detail in Chapter 20.

Trash in the line

Low water pressure may be caused by obstructions in the water pipes. When new plumbing is installed, it is subjected to various conditions that may cause trash to find its way into the pipe. The foreign material may be dirt, wood chips, or some other undesirable material. Dirt is probably the most likely candidate for the obstruction. If the end of a water pipe is set on the ground during the installation, it can pick up soil. This type of blockage may very well go unnoticed during the installation.

After the water is turned on to the dirt-filled pipe, the dirt is forced to a stopping point. Depending on the amount of debris, it may stop at a supply tube or the faucet's aerator. In either event, the water pressure will not be what you expect. This type of problem will usually give you some hints of its existence. You will see dingy water coming from the faucet. If you have low water pressure and dingy water, check for partially blocked supply tubes and aerators.

Water conditioners

Water conditioners can contribute to low water pressure. These water treatment systems can be ignored in the standard troubleshooting process. Since the potable water must pass through the water treatment center, you could have low water pressure if the treatment equipment is not installed properly. There are many different styles of water conditioners. Refer to the owner's manual for ways to check the performance of the water treatment system.

Old piping

If you have connected the new plumbing to existing water pipes, the old pipes may be responsible for the low water pressure. Galvanized steel pipes especially are likely to contribute to low water pressure. Mineral deposits, rust, and other substances build up on the inside of the pipe and restrict the flow of water. When you tie into one of these restricted pipes, the resulting water pressure might be low.

This cause of low water pressure is ignored by many plumbers. They assume that if the existing pipes are causing trouble, low water pressure would have been noticed before the new installation. It is a reasonable assumption, but it is not always correct. There are fixtures in the home where low water pressure is rarely noticed. For example, the pipes that supply water to the washing machine may produce low pressure that goes unnoticed. The water pressure filling a toilet usually is not noticed. However, if you tap into these pipes to supply a lavatory faucet with water, the low pressure will be evident. Don't rule out the possibility of old pipes being the cause of the low pressure, even if the low pressure is not found throughout the house.

Noisy pipes

Many houses suffer from the irritating noise of banging water pipes. The pipes in these houses can make some serious noise when the water is used. There are two common causes for noisy water pipes: a water hammer and a lack of support for the pipes. Correcting these problems after the plumbing is installed can be perplexing. To correct either of these causes, you must have access to the piping. After the plumbing is installed and the walls are covered, getting to the pipes might be a problem.

Poor pipe support

The plumbing code requires that water pipes be supported at intervals of 6 feet. However, not all pipes are installed in strict code compliance. Many plumbers fail to install adequate support mechanisms for water pipes. When the pipes are not secured at the prescribed intervals, they can vibrate when the water surges through them. The vibration can cause the pipe to be noisy; also, it eventually might wear a hole in the pipe.

If the pipes rattle when the water is used, the noise is probably due to poor pipe support. If you can access the pipes, adding additional support should correct the problem. If the pipes are concealed in walls and ceilings, adding new hangers will not be easy. You must assess the situation and determine your willingness to correct the problem. In such cases, you will have to cut open the walls or ceilings to gain access to the pipes. If you are not willing to open the walls or ceilings, you will be forced to live with the rattle of pipes every time you use the plumbing.

In time, the constant vibration may wear holes in the pipes. When the vibration of the pipes causes a hole, water will begin to spray inside the walls or ceilings. Then you will have no choice. You will have to open the area and repair the pipes. The best way to solve the agony of rattling pipes is to support them well when they are installed.

Water hammer

If the pipes make a loud banging sound, you are the victim of a water hammer. In some instances, the banging of the pipes can sound like a shot from a small-caliber gun. A water hammer does not usually affect every fixture in the home. It is most prevalent on fixtures with quick-closing valves. Two examples of this type of fixture are the toilet and the washing machine. A water hammer may be present at faucets, but it is most common at quick-closing valves.

What causes a water hammer? A water hammer is caused when fast-moving water is stopped suddenly. When the ballcock in the toilet shuts off the water, it does so very quickly and encourages a water hammer. A water hammer is more likely if you have installed the pipes in a long, straight run.

For example, if you have installed 30 feet of straight pipe and then turned it up on a 90-degree angle to go to a faucet, you may develop a water hammer. The water rushes down the long straight run until it gets to the 90-degree turn. When the water hits the back of the elbow fitting, it can cause the pipe to bang. A water hammer more likely will occur when there is strong water pressure in the plumbing system.

How do you eliminate a water hammer? The best way to eliminate a water hammer is to plan for it in the plumbing installation. Avoid long runs of straight pipe. To do this, offset the pipes at various intervals to slow down the water's speed. When you put in new water pipes, install water hammer arrestors. You may buy these arrestors or you may make your own. To make your own, use a 12-inch length of pipe and a cap. The pipe should be larger than the pipe bringing water to the fixture. For example, if you are running a 1/2-inch pipe to the toilet, the water hammer tube should be made 3/4- or 1-inch pipe.

These water hammer tubes and arrestors may be installed after the initial plumbing if you have access to the supply pipes. The water hammer tubes should be installed above the cut-off valve for the fixture. Refer to Fig. 19-3 for an example of how a typical water hammer tube, or air chamber, may be installed. Installing these precautionary devices during the initial plumbing installation is easy and not very expensive. Installing them after the job is finished is more challenging. If the job is finished and you have a water hammer, here are a few ways you might be able to solve the problem without tearing into the walls.

Eliminate long, straight runs of pipe. If you have a basement or a crawlspace, you may be able to access the primary water pipes. If you can, look at these pipes to see if they run long distances without being offset. If they do, alter the piping to incorporate offsets. Do this by using elbow fittings to turn the pipe out and back in again.

Install air chambers on the primary water pipes. If you have access to the primary water pipes in the basement or crawl space, adding air chambers may reduce or eliminate the water hammer. Cut a tee into the primary water pipe. In the branch fitting, install the air chamber so that it rises as high as possible above the primary water pipe, up to 18 inches. The installation of these air chambers in the primary water pipes at each location where a branch pipe leaves to serve a fixture should reduce the water hammer. Add air chambers at the water heater, the water main, and anywhere else possible. This is not the ideal way to install air chambers, but it is better than not having them at all.

Fig. 19-3. Sillcock with an air chamber installed.

Install air chambers at each fixture. Unfortunately, when you add air chambers after the job has been completed and the pipes concealed, you will not be able to make a proper installation. You may still be able to make an installation with a compromise. Look under the sinks to see how much room you have to work. If the supply pipes come out of the wall, you may be able to cut a tee in between the wall and the cut-off valve. If you can, cut in a tee and run the air chamber up as high as you can. If the supply pipes come up through the floor, cut in a tee, come out of the branch opening, and turn up with an elbow. From the elbow, extend the air chamber as high as possible. These methods are not ideal, but they will help to reduce the water hammer.

Other problems

If you are having a problem that has not been covered in this chapter, look to the next two chapters. Chapter 20 deals with problems you may have with faucets and fixtures. Chapter 21 is dedicated to finding and fixing pump problems. If the solution to the water pipe problem was not covered in this chapter, it is probably in one of the next two chapters.

20

Faucet and fixture leaks and failures

AFTER ALL THE PLUMBING is installed and tested, you may still have some work to do. It is not uncommon for new plumbing installations to develop leaks in the first few days of operation. Neither is it unusual for new fixtures and faucets to need some type of adjustment in the early weeks of their usage. The owner's manuals and instruction sheets that are packed with new equipment do not cover all the possible problems you may encounter. This chapter picks up where the installation instructions end. The tips in this chapter will help you to troubleshoot and correct deficiencies in newly installed plumbing. They will also be useful for many existing plumbing repair situations.

Relief valves on water heaters

The relief valve on the water heater is a very important safety device. The relief valve protects you when a water heater develops excessive temperature or pressure. Without a dependable relief valve, the water heater can turn into a bomb. Never take the relief valve for granted.

After a new installation, relief valves can develop drips and leaks. If the valve is leaking around its threads where it is screwed into the water heater, you must remove and reinstall the valve. The cause of this type of leak is usually a lack of dope or teflon tape. If the valve drips out of its discharge tube, you have to decide why it's leaking.

The leak might be caused by sediment in the relief valve. If debris gets between the valve's seat and washer, drips are going to happen. To test this, open the valve and allow it to blow out some water. Then close the valve and see if it seals. Sometimes letting the relief valve blow off some water will clear the debris. If the valve continues to drip, replace it.

When the relief valve blows off at full capacity, you might have a bad valve. The other possibility is thermal expansion. If the water distribution system is equipped with a backflow preventer, thermal expansion can force the relief valve to blow off. Change the relief valve and observe it for several days, or until it blows off again. If the valve operates normally, you have corrected the problem. If the new valve blows off, you either have thermal expansion or a problem with the water heater.

If you have a backflow preventer, thermal expansion is a good guess. The reason backflow preventers cause thermal expansion is their unwillingness to allow expanding water to run back down the water service pipe. Without a backflow preventer, expanding water can push back down the water service and into the water main. The backflow preventer blocks the path and does not allow water to flow back out of the house. When the faucets are closed, there is no place for the expanding water to go. The excessive pressure forces the relief valve to blow off. To correct the problem, you must provide a place for the expanding water to go.

To reduce the effects of thermal expansion, you may install air chambers on the water pipes at the water heater. In many cases, the installation of air chambers will resolve the problem. If the air chambers don't get the job done, you may have to install an expansion tank. The expansion tank will be similar to the type used on boilers. As the thermal expansion occurs, the expanding water escapes into the expansion tank.

Water heater failures

When the water heater is not up to par, you may have limited or no hot water. Working with water heaters is potentially dangerous. Electric water heaters have high voltage around all of the interior parts that might fail. These parts are the heating elements and thermostats. With gas water heaters, there is the risk of fire or a gas explosion. Oil-fired water heaters also have the potential for fire-related risks.

Since most homeowners and novice plumbers are not experienced with these dangerous elements, I am not comfortable giving you advice on how to work on the more dangerous parts of the water heater. While I do not advise you to attempt repairs on these items, I will give you a broad view of what might be wrong with the water heater when it fails so you can understand the service technician's repairs.

Electric water heaters

Heating elements. If you have a limited supply of hot water, which diminishes quickly, you probably have one bad heating element. Water heaters in most homes have two heating elements. Some heating elements use bolts and others screws (Figs. 20-1 and 20-2). If both elements go bad, you will not have any hot water. To determine if the elements are bad, a certified technician must test them with a meter while the electricity is turned on. This involves working directly with 240 volts of electricity. Do not attempt to test the heating elements. The risk of fatal electrocution is too great for the average person.

Fig. 20-1. Screw-in heating
element for electric water heater.

Fig. 20-2. Bolt-in heating element
for electric water heater.

Thermostats. Electric water heaters have thermostats that may go bad (Fig. 20-3). Most home-installed water heaters have two thermostats, an upper and a lower. Again, electrical power is present at the thermostats. The thermostats are set into a bed of insulation, behind the access panels of the water heater. The wires that provide the power are also hidden in the insulation. Only trained technicians should remove the access covers on electric water heaters. The high voltage is sitting just behind the panels, waiting to zap untrained hands. If you get hit by the power from these wires, you could be killed.

Reset buttons. There are reset buttons on electric water heaters. These buttons are located behind the access panels, near the electrical wires. I don't

Fig. 20-3. Thermostat for electric water heater.

believe the average consumer should attempt to use these reset devices. The danger of working around the high voltage is just too great. If you insist on resetting the water heater, refer to the owner's manual for instructions.

Water temperature adjustments. There is a dial located behind the access cover of an electric water heater that allows you to set the temperature of the water. This dial may be set when the electrical power is turned off. The dial is normally set at the factory, and is sometimes adjusted when the water heater is installed. If you decide to change the temperature setting on the water heater, cut off the power to the water heater first.

Even when the power is turned off, be very careful. Many times when someone thinks the power is off, it isn't. Turn the dial to the right or left to raise or lower the water temperature. This is the only work behind the access cover that the average homeowner should attempt. I still do not recommend that you do this job. If the power to the water heater is not off, you might brush against a hot wire. Just because a label on a circuit breaker indicates it controls the power for the water heater, does not necessarily make it so.

Gas and oil-fired water heaters

Gas water heaters have a dial on the outside that controls the water temperature. There is also a dial that controls the flow of gas. Beyond setting the water temperature control, there is nothing else on a gas water heater that anyone, except a professional, should do. When it comes to the gas lines and burner controls, leave them to the experts. This same advice holds true for oil-fired water heaters. Other than setting the water temperature, there is nothing the average person should work with on an oil-fired water heater.

Frost-free sillcocks

Frost-free sillcocks do not require much attention. If sediment gets between the seat and the washer, the sillcock might drip. The packing nut on the sillcock might be loose and leaking. These are the only two problems you are likely to experience. If the packing nut is leaking, all you have to do is tighten it. Unless the nut is cracked or the packing is defective, tightening it will be all that is necessary.

If water is dripping out of the sillcock, you will have to remove the stem. Cut off the water to the supply pipe. Look at the handle of the sillcock and you will see a nut that holds the stem inside the body. Loosen the nut and the stem will come out. Inspect the washer on the end of the stem. If the washer is good, leave it alone. If it is cut or worn, replace it. While the stem is out, cut on the water to the pipe that supplies the sillcock. Water will blow out of the end of the body and flush any sediment and debris from the seat. After a few moments, cut off the water. Put the stem back into the body and tighten the retainer nut. Cut on the water and inspect the results of your work. This procedure should cure the drip.

Washing machine hook-ups

Washing machine hook-ups may be nothing more than two boiler drains, or they may consist of a faucet device. In either case, there is not much call for adjustments. The boiler drains have a packing nut that might leak. They might also drip from a bad washer if debris gets between the washer and the seat. To correct either of these problems, follow the directions given in the section on sillcocks.

If the hoses connect to a faucet, there will be little you can do if the faucet is defective. This type of faucet is inexpensive and has very few repairable parts. In most cases, if the faucet is bad, replace it. If you have problems with the water pressure, check the cone-shaped screens in the ends of the hoses. These screens filter sediment from the water pipes. In new installations, it is not uncommon for these screens to become clogged.

Tub and shower faucets

When you must work on a faucet, refer to the installation instructions that came with the faucet. With the number of different faucets available, it is difficult to give you precise instructions. Instead, I will give you instructions for the most common types.

Two-handle faucets

Two-handle faucets may be used for either bathtubs, showers, or a combination of the two. When two-handle faucets are used for tub and shower combinations, the tub spout will have a diverter on it (Fig. 20-4). This is the little rod that you pull up to divert water from the tub spout to the shower head. When the diverter rod is lifted, the bulk of the water should come out of the shower head. If you lift the rod and still have a substantial flow of water coming from the tub spout, you need a new diverter. This requires that you replace the existing tub spout with a new one.

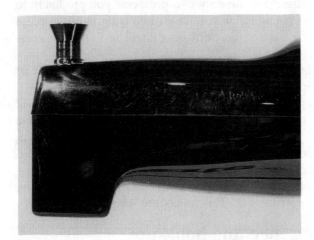

Fig. 20-4. Tub spout with diverter.

If water drips from the tub spout after the faucets are cut off, you either have a bad washer, a bad seat, or debris between the washer and the seat. To correct these problems, you will need a set of tub wrenches (Fig. 20-5). These deep-set wrenches allow you to reach inside the wall to turn the nuts that hold the stem into the faucet body. Before you remove the faucet stems, cut off the water to the faucet body.

The handles of the faucets will be attached with screws. The screws are hidden by caps that pop off. Use a knife to pry the caps out of the center of the handles. Loosen the screws and remove the handles. Then remove the trim escutcheons. Most of these may be removed by turning them counterclockwise. In some cases, the escutcheons will be held in place with a set screw. If this is the case, you will need a hex wrench to loosen the set screw. Slide the tub wrench over the stem and into the wall. When the wrench is on the retainer nut, turn the wrench counterclockwise. You can use tongue-and-groove (t&g) pliers or a pipe wrench to turn the tub wrench. Once the nut is loose, the stem will come out.

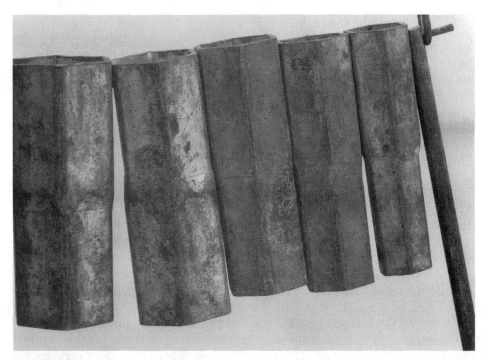

Fig. 20-5. Tub wrenches.

Inspect the washer and replace it if it appears damaged. If you suspect that there is debris in the faucet body, flush it out. Put the faucet back together and test it. If it still leaks, you may have to replace the seat. Cut off the water and remove the stem. Insert a seat wrench into the seat and turn it counterclockwise. After a few turns, the seat will come out. Replace it and put the faucet back together. If all these efforts fail, either call a professional or replace the faucet.

If water is leaking around the handles of the faucet, you must either tighten the packing nuts, replace the packing, or replace the "O" rings. To do either of these jobs, remove the handles from the stems. If you are only tightening the packing nuts, you do not have to cut off the water. If you plan to replace the "O" rings, you must cut off the water. The packing nuts are the nuts around the stem. You will see water leaking past the nut if it needs to be tightened. Turn the packing nut clockwise to stop the leak in a new faucet. In an old faucet, you may have to replace the packing material.

If you must replace the packing material, cut off the water. Loosen the packing nut and slide it back on the stem. Wrap the packing material around the stem with a greasy string and replace the packing nut. The nut will compress the packing and hold it in place.

The "O" rings are located on the outside of the faucet stem. Remove the stem from the faucet to get to the "O" rings. If the rings are damaged, replace them. Then reinstall the stem.

Three-handle faucets

Three-handle faucets are the same as two-handle faucets, except for the diverter. Instead of having a diverter on the tub spout, the center handle diverts the water from the tub spout to the shower head. If the water is not diverting properly, either replace the washer or the seat. In new installations, you may have to clear debris from within the body of the diverter. The techniques for this work are the same as described above.

Single-handle faucets

There are two basic types of single-handle faucets. They are the cartridge-type and the ball-type. While the two types may look the same on the fixture, there are significant differences on the internal parts of the faucets. With either type of faucet, cut off the water before you repair them. In both types, you must remove the handle from the stem. This is where the similarities end.

Cartridge-style faucets. Before you proceed, be sure the water is off. Remove the trim collar that sits over the stem. This collar pulls up for removal. You should see a thin clip that holds the stem inside the faucet body. Use a pair of pliers to remove this clip. Grip the clip on the tab extending from the body and pull it straight out. Once the clip is out, grasp the stem with the pliers and pull it out. The cartridge will have "O" rings on the outside of it. These cartridges are made to be replaced, not repaired. The only exception is the "O" rings. If the "O" rings are bad, replace them.

Ball-type faucets. The ball-type faucet has many parts. The handle on this type of faucet is usually held in place with a screw. Remove the screw and lift the handle off the stem. There should be a large, knurled nut holding the ball in place. You may have to use pliers to loosen this nut, but be careful not to scratch its finish.

When the large nut is removed, you will see a ball sitting in the faucet. It will have "O" rings, springs, and other loose parts fitted into it. Remove the ball carefully. You may purchase repair kits for these balls, or you may replace the entire assembly. Refer to the installation manual for exact instructions on how to repair or replace the ball in the faucet.

Lavatory faucets

Most lavatory faucets have a single body with one or two handles. The repair and replacement procedures for the internal parts are essentially the same as those described above. The only big difference is that lavatory faucets are easier to repair. You will not need deep-set wrenches to reach the nuts on a lavatory faucet. Once the stem escutcheons are removed, you can loosen the retainer nuts with an adjustable wrench.

Kitchen faucets

Kitchen faucets are about the same as lavatory faucets when it comes to repairs. The only major difference is the spray hose assembly, if the kitchen faucet has one. If the spray is not working well, it may be clogged with sediment. If the water does not divert from the kitchen spout to the spray as it should, the handle of the spray may be restricted. You may unscrew the spray head and attempt to clean it, or you may replace the entire head.

Toilets

It is very common for new toilets to require some adjustments after installation. The adjustments needed will usually be in the toilet tank.

Ballcocks

There are two common types of ballcocks. The first has a horizontal float rod with a float or ball on the end of it. The other type has a plastic float that moves up and down the shaft of the ballcock. The ballcock controls the amount of water that enters the toilet tank. Ballcocks are adjustable to increase or reduce the amount of water in the tank. Ideally, the tank's water level should be level with the fill line etched into the interior of the tank.

The float-rod ballcock is very common and is easily adjusted (Fig. 20-6). If you want more water in the tank, bend the float rod down so more water is required to float the ball high enough to reach the cutoff point. If you want less water, bend the float rod up. This causes the ball to reach its cutoff point earlier, with less water. The float rods are easy to bend with your hands.

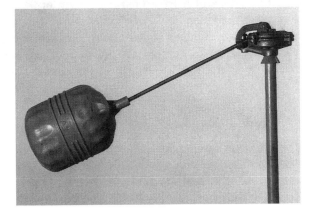

Fig. 20-6. Horizontal, float-type ballcock.

To bend the rod, place the palm of your hand either over or under the rod, depending upon which direction you wish to bend it. With the rod cradled in the palm of one hand, use your other hand to bend it at the end with the float. Be careful not to bend the rod sideways. If the float gets too close to the side of the tank, it may stick on the tank.

As float-type ballcocks age, the float rod often becomes loose. When the rod is not screwed into the ballcock securely, the float is apt to come in contact with the side of the tank. When the float is hung up on the tank, the ballcock will not work properly. If the ball sticks in the down position, the water will not cut off. If it sticks in the up position, the water will not cut on.

Vertical ballcocks have not been around as long as float-rod ballcocks, but they are very popular (Fig. 20-7). One of the big advantages to a vertical ballcock is that the float can't stick to the side of the tank.

To control the water flow with a vertical ballcock, move the float up and down the shaft of the ballcock. Moving it down produces more water; moving it up produces less. To move the float, squeeze the thin metal clip that is located on the metal rod next to the float. While you squeeze the clip, move the float up or down. When you like the placement, release the clip and the float will remain in the selected position.

Flappers

If the flush valve has a flapper, it may need to be adjusted. If the chain between the flapper and the toilet handle is too short, the flapper will not seal the flush hole. This allows water to run out of the tank and into the bowl. As the water level drops, the ballcock refills the tank. This is an endless cycle that wastes a lot of water. If you notice the ballcock cutting on at odd times or see water running into the bowl long after a flush, investigate the flapper's seal. Move the chain to the next hole in the handle to improve the seal.

If this doesn't help, adjust the length of the chain where it connects to the handle. If the chain between the flapper and the handle is too long, the flapper may become entangled in the chain. This causes the flapper to be held up and allows water to constantly run down the flush hole. This problem can be corrected by shortening the length of the chain. The point is, make sure that the flapper sits firmly on the flush hole.

Tank balls

Tank balls are sometimes used instead of flappers. If you have a tank ball, you also have lift wires and a lift-wire guide. The guide is attached to the refill tube in the toilet tank. The lift rods attach to the ball, run through the guide, and attach to the toilet handle. If the alignment of the tank ball is not set properly, water will leak past it and into the toilet bowl.

If you are having problems with the tank ball, inspect the location of the ball on the flush hole. If it is not properly seated, adjust the lift wires or the guide to gain a satisfactory seal. The guide can usually be turned with your fingers to

Fig. 20-7. Vertical ballcock.

make a better alignment. If the lift wires are too short and hold the ball up, you must bend them to add length.

If these attempts do not prove fruitful, you may have to replace the tank ball. Unscrew the ball from the lift rod and replace it with a new ball. If a new ball doesn't do the trick, the flush valve may be pitted or defective. Use sandpaper on a brass flush valve to remove pits in the brass. Sand the area where the tank ball sits to smooth out the rough spots and encourage a better seal. The pits are voids that a tank ball cannot seal. They allow water to leak through to the toilet bowl. Unless you are working with an old toilet, the flush valve should not be pitted.

Tank-to-bowl bolts

As a new toilet is used, the tank-to-bowl bolts might work loose. This is especially true if the tank is not supported by a wall. If you develop leaks at the tank-to-bowl bolts, tightening the bolts should stop the leaks. Be careful not to tighten the bolts too tight. If you do, the china will break.

The flush-valve gasket

If you have water flooding out from between the tank and bowl when the toilet is flushed, you have a poor seal at the flush-valve gasket. This is the grey sponge or black rubber gasket you fitted over the flush valve when you installed the toilet. If the tank-to-bowl bolts are not tight enough to compress this gasket, water will leak each time the toilet is flushed. Normally, tightening the tank-to-bowl bolts will stop this type of leak. If tightening the tank-to-bowl bolts doesn't correct the problem, you will have to disassemble the toilet and further investigate the problem.

It is possible that the flush-valve nut is cracked. It is also possible that the threaded portion of the flush valve has a hole in it. The gasket may have been shifted during the installation, rendering it ineffective. A good visual inspection should reveal the cause of the leak. After making the necessary corrections, reassemble the toilet and test it. If you have a continual problem, try replacing the flush valve.

Garbage disposers

The biggest problem with garbage disposers is that they become jammed. If the disposer hums, but does not turn, the cutting heads are stuck. To correct this problem, cut off the power to the disposer. Then place a broom handle into the disposer and wedge it against the jammed cutting blades. With a little forceful leverage, you should be able to free the blades. Some disposers come with a special wrench that is designed to clear the cutting heads. Refer to the owner's instructions to get the disposer back into service.

If the disposer fails to start, you may try pushing the reset button. Most disposers have a reset button on the bottom of the unit. The owner's literature should detail the location of the reset button. If pushing the reset button doesn't correct the problem, you may have a blown fuse or circuit breaker. If there are no blown fuses or breakers in the electrical panel, call in a professional.

Expect the unexpected

After successfully installing and testing a new plumbing installation, many people assume the job is over. Unfortunately, even professionals have to make adjustments and correct leaks after the job apparently is done. When small leaks go unnoticed for a long time, they can cause sizable damage. It will be in your best interest to catch and correct any deficiencies in the system early.

21

Pump problems

WHEN THE POTABLE WATER comes from a well, you are dependent on the water pump. If the pump fails, you will not have any water. Many factors might cause the pump to malfunction. You might have an electrical problem. The pump controls might be bad. The pressure tank might be waterlogged and causing the pump to run too frequently. There are dozens of ways for you to be without water when you have a pump. This chapter will show you how to troubleshoot and repair many of these problems.

Many times a pump's failure is related to the electrical system. I don't recommend that you work with or around electricity without proper training. Much of the troubleshooting data in this chapter involves electricity. If you are not experienced in electrical work, do not attempt these procedures. When in doubt, call in a professional plumber or electrician.

The following section will explore general items to check if the pump will not run, runs but doesn't produce water, or cuts on or off too frequently. Items specific to jet or submersible pumps will be addressed later in this chapter.

When the pump will not run

When the pump won't run, there are six likely causes for the malfunction: a tripped circuit breaker, damaged or loose wiring, a defective pressure switch, stopped-up nipples, voltage problems, or a defective motor.

Circuit breakers and fuses

The first item to check is the fuse or circuit breaker that controls the pump's electrical circuit (Fig. 21-1). If the fuse is blown or the breaker is tripped, the pump will not run. When you check the electrical panel and find these conditions, replace the blown fuse or reset the tripped breaker. In future references to fuses or circuit breakers, I will refer only to circuit breakers. If the home has fuses, when I indicate the need to reset the circuit breaker, you should replace the blown fuse.

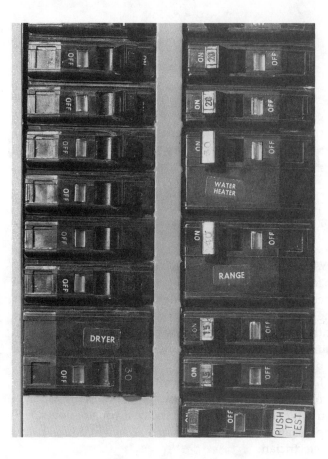

Fig. 21-1. Circuit breaker panel.

Damaged or loose wiring

If the circuit breaker is not tripped, inspect all the electrical wiring that affects the pump. There may be a broken or loose wire that is preventing electricity from reaching the pump. Remember you are dealing with electricity. Use appropriate caution to avoid electrical shocks.

Pressure switches

The pressure switch might be defective (Fig. 21-2). It is also possible that the switch is out of adjustment. To correct this problem, remove the cover on the pressure switch. You will see two nuts sitting on top of springs. The nut on the short spring is set at the factory and should not need any adjustment. However, if you want the pump to cut off at a higher pressure, turn this nut clockwise. To have the pump cut off at a lower pressure, turn the nut counterclockwise. The nut on the taller spring controls the cut-on and cut-off cycle of the pump. To make the pump cut on at a higher pressure, turn the nut clockwise. If you want the pump to cut on at a lower pressure, turn the nut counterclockwise.

Fig. 21-2. Pressure switch, control box, and electrical disconnect.

Stopped-up nipples

If the tubing or nipples on the pressure switch get clogged, the pump might not run. If you suspect this to be the problem, take the pipe and fittings apart and clean or replace them. This problem is unlikely, but possible, in a new installation.

Voltage problems

If the pump is not receiving the proper voltage, it will not run. This problem calls for direct contact with the electrical system. Due to the danger involved with electricity, I will not help you in this aspect of troubleshooting. If you have eliminated all the non-electrical problems, call in a professional plumber or electrician to fix the pump.

Seized pumps

If the pump becomes mechanically bound, it will not run. To correct this problem, remove the end cap from the pump. Once the end cap is removed, you should be able to turn the motor shaft with your hands. Once the shaft rotates freely, reassemble the pump and test it. The problem should be solved.

Overloaded motors

If the motor becomes overloaded, the protection contacts will remain open. This problem will solve itself because the contacts will close automatically after a short time.

Defective motors

Obviously, a defective motor will prevent the pump from running. To determine if the pump's motor is bad, you must do extensive electrical troubleshooting. You will have to check the ground, capacitor, switch, overload protector, and winding continuity. If you have experience working with electrical systems, you already know how to do this. If you do not, I will not give you instructions. One wrong move could deliver a potent electrical shock. Again, if the problem is electrical, call in a professional.

The pump runs, but produces no water

Several factors may keep an operating pump from supplying water. Troubleshooting techniques for the most common problems are given below. Do not allow a pump to operate if it is not producing water because the motor might be damaged.

Pressure control valves

In a two-pipe system, the pressure control valve might need to be adjusted when the pump runs without producing water. When the pressure control valve is set too high, the air volume control might not function. When the pressure control valve is set too low, the pump might not cut off. To reduce the water pressure, turn the adjusting screw on the pressure reducing valve counterclockwise. To increase pressure, turn the adjusting screw clockwise.

Water level

If the well pipe is not below the water level, the pump cannot produce water. With shallow wells, it is not uncommon for the water level to drop during hot, dry months. If the pump is running, but not producing water, check the water level in the well. You can do this with a roll of string and a weight. Tie the weight onto the end of the string and lower it into the well. When the weight hits bottom, withdraw the string. The end of the string should be wet. Measure the distance along the wet string. This will tell you the depth of the water in the well. This information, along with the records of how long the well pipe is, will tell you if the end of the pipe is submerged in the water.

When a deep well is drilled, the installer should provide you with the recovery rate of the well. The recovery rate will be stated in gallons per minute (gpm). The minimal recovery rate normally accepted is 3 gpm. A better rate is 5 gpm. If the pump is pumping faster than the well can recover, you may run out of water. Pumps are rated to tell you how many gallons per minute they pump. Compare the pump's rating to the recovery rate of the well and you may find the cause of the trouble.

Clogged foot valves

If the foot valve is too close to the bottom of the well, it may become clogged with mud, sand, or sediment. It is not uncommon for shallow wells to fill with sediment. What starts out as a 30-foot well may become a 25-foot well as the earth settles and shifts. Even if the foot valve was originally set at the appropriate height, it may now be hanging too low in the well. Sometimes shaking the well pipe will clear the foot valve. If this doesn't work, pull the pipe out of the well and clean or replace the foot valve. If the foot valve is pulling up sediment from the bottom of the well, shorten the well pipe.

Malfunctioning foot valves

Foot valves act as a check valve (Fig. 21-3). If the foot valve or check valve is stuck in the closed position, you will not get water from the pump. Pull the well line and inspect the check valve or foot valve. If the valve is stuck, replace it. Sometimes a check valve will be installed backward. If you are experiencing problems with a new installation, check to see that the check valve is installed properly.

Goulds Pumps, Inc.

Fig. 21-3. Foot valve.

Defective air-volume control

It is possible that the diaphragm in the air-volume control has a hole in it (Fig. 21-4). Disconnect the tubing and plug the connection in the pump to determine if there is a hole in the diaphragm of the air-volume control. If plugging the hole corrects the problem, replace the air-volume control.

Goulds Pumps, Inc.

Fig. 21-4. Air-volume control.

Suction leaks

A common ailment of shallow-well systems is a leak in the suction pipe. To check for this problem, pressurize the system and inspect it for leaks. If you find a leak, fix it, and you will have water again.

When the pump will not cut off

If the pump builds pressure, but will not cut off, you might have a bad pressure switch. Before you replace the switch, check a few other things. Check all the tubing, nipples, and fittings associated with the switch. If they are obstructed, clear them and test the pump. If the problem persists, check the pressure switch to see that it is set properly. Do this by checking the cut-on and cut-off controls discussed earlier. If all else fails, replace the pressure switch.

When you don't get enough pressure

When the air-volume control is on the blink, you might not be able to build suitable pressure. Test the control as described above and replace it, if necessary. On older pumps, the impeller hub or guide vane bore may be worn. If this is the case, you will need to call a professional to verify the diagnosis. If either of these parts are worn, replace them.

When the pump cycles too often

Leaks in the piping might cause the pump to cut on and off more frequently than it should. When the system is pressurized, check all piping for leaks. If any are found, repair them. If the pressure switch is not working correctly, the pump may run at random. Follow the directions given earlier to troubleshoot the pressure switch. If necessary, replace the switch.

The problem might be in the suction lift of the system. If water floods the pump through the suction line, control the water flow with a partially closed valve. You also might not be developing enough vacuum on the suction line to get the job done. During the troubleshooting process, check out the air-volume control. It is possible the valve is defective. The instructions given earlier will explain how to check the air-volume control.

Pressure tanks

The most likely cause when pumps cycle too often is a faulty pressure tank (Fig. 21-5). If the tank has a leak in it, it cannot hold pressure. This is a common problem in old tanks. If you find a leak in the tank, replace the tank. If the air pressure in the pressure tank is not right, the pump will cut on frequently. The pump may cut on every time the water is turned on to a faucet. This condition should not be ignored; it could burn out the pump motor.

Modern captive air tanks come with a factory pre-charge of air. These tanks use a diaphragm to control the air volume. Older tanks do not have these diaphragms. It is not uncommon for older tanks to become waterlogged. When the tank is waterlogged, it contains too much water and not enough air.

Fig. 21-5. Diaphragm-type pressure tank. Goulds Pumps, Inc.

Checking the tank's air pressure

To check the pressure tank's air pressure, cut off the electrical power to the pump. Drain the tank until the pressure gauge is at zero. A rule-of-thumb for the proper air charge dictates that the air pressure should be 2 pounds per square inch (psi) less than the cut-on pressure of the pressure switch. For example, if the pressure switch is set to cut on at 40 psi, the air in the tank should be set at 38 psi. You can check the air pressure by putting a tire gauge on the air valve of the tank.

If the air pressure is too low, pump air into the tank. You can do this with a bicycle pump or an air compressor. Monitor the pressure as you fill the tank with air. When the air pressure is at the proper setting, turn on the electrical power to the pump. When the water floods the tank, you should have a normal reading on the pressure gauge. The system should work properly if the air pressure remains at the proper setting.

If the pressure tank is old, it may have a small leak that allows air to escape. It is possible that a new tank is defective. If this procedure does not keep the system working, consider replacing the pressure tank.

Jet pump problems

A loss of the pump's prime

If the pump loses its prime, the pump will run without producing water. With a shallow-well pump, remove the priming plug and pour water into the pump. When the water level stands static at the opening, apply pipe dope to the priming plug and replace it. Start the pump and you should get water pressure.

If you have a two-pipe system, disconnect the pipes from the pump. Pour water down each pipe until they are both full of water and the water level holds. Then reconnect the pipes to the pump. Prime the pump through the priming hole. Start the pump and you should get water. If you don't, the problem may be in the pressure control valve.

Failing jet assemblies

If the jet assembly fails, the pump will not produce water. Inspect the jet assembly for possible obstructions. If an obstruction is found, clear it. If you cannot clean the assembly, replace it.

Lifting water too high

If you try to lift water too high, the pump may not be able to handle the job. Check the rating of the pump to be sure it is capable of performing under the given conditions. If you have leaks in the piping, the pump will not be able to build adequate pressure. Check all piping to confirm that no leaks are present. If the jet or the foot valve is partially obstructed, the water pressure will suffer. Check these items to be sure they are free to operate. If they are blocked, clear the obstructions and test the pump.

Submersible pump problems

Control box

If the pump has never run properly, you may have the wrong control box (Fig. 21-6). Inspect the box and confirm that it is the correct box for the pump. When the control box is exposed to high temperatures, you may experience trouble with the pump. If the control box is located in an area where the temperature exceeds 120 degrees Fahrenheit, the high temperature may be the cause of the problem.

Bad motor

If the pump motor is badly worn, it may not produce water even if it runs. Pull the pump and inspect the motor. If it is worn, replace it.

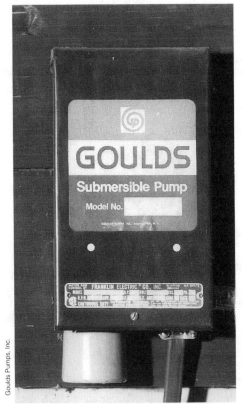

Fig. 21-6. Control box.

Clogs

If the protective screening on the pump is obstructed, water cannot enter the pump. This will cause the pump to run without providing water. If the impellers are blocked by obstructions, you will experience the same problem. Pull the pump and check the strainer and impellers for clogs.

Leaks

If there are leaks in the piping, the water may not be able to get to the house. If you can connect an air compressor to the pipe, pump it full of air to test for leaks. Inspect all the piping for leaks and repair or replace the pipe as necessary.

Waste-water pumps

Waste-water pumps include sump pumps, sewage ejectors, laundry-sink pumps, and all other pumps that pump water not intended for drinking. In this section, I will give you information on how to troubleshoot and repair several of these pumps.

Vertical sump pumps

Vertical sump pumps are often used in wet basements (Fig. 21-7). The motor for these pumps is mounted on top of the pump's shaft. The motor is not submerged in the water. There is a float that moves up and down and controls when the pump cuts on and off. Most of these pumps have an electrical cord that plugs into a standard outlet.

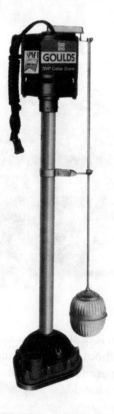

Fig. 21-7. Vertical sump pump.
Goulds Pumps, Inc.

If the pump will not pump water, check the circuit breaker that controls the pump. If the breaker is not tripped, check to see if the float is stuck. If the float is stuck in the down position, the pump will not run. Stuck floats are a common problem with sump pumps. Another possible cause for the problem is the check valve in the discharge pipe. If the check valve is installed backwards or is stuck in the closed position, the pump will not work properly. If the pump still does not work, check the strainer on the bottom of the pump. If debris is blocking the strainer or impellers, the pump will not function properly.

Submersible sump pumps

Submersible sump pumps operate below the water level (Fig. 21-8). Their floats may be internal or external. If you have an external float, it is not uncom-

Fig. 21-8. Submersible sump pump. Goulds Pumps, Inc.

mon for it to become stuck. The suggestions given for troubleshooting vertical sump pumps will apply to submersible pumps as well.

If the pump will not cut off, check the float to see if it is stuck in the up position. If the check valve is stuck in the open position, the pump will run on a regular cycle. After it pumps water out of the basin, the pump will shut off. When the pump shuts off, the water in the vertical and back-graded discharge pipe will run back into the basin. This will cause the pump to cut on and pump the same water over and over. This condition will eventually ruin the pump. To correct the problem, repair or replace the check valve.

Most sump pumps have a cord that is plugged into a standard outlet. There are switches and electrical aspects of the pump that might go bad. Since these problems are electrical, you should call a professional to correct them, unless you are accomplished at electrical work.

Sewer pumps

Sewer pumps work on a principle similar to sump pumps (Fig. 21-9). Follow the same troubleshooting advice given above for problems with sump pumps. If you have an electrical problem or a problem not covered in these tips, call a professional.

Laundry pumps

There are different types of laundry pumps. One type sits in the basin that collects the drainage from the laundry tub. When the water level in the basin raises the pump's float, the pump empties the basin. This type of pump works like a submersible sump pump.

The other type of pump attaches to the drain of the laundry tub and pumps the water directly from the laundry tub. This type of pump may cut on automatically when the water reaches a certain level, or it may need to be operated manually with a switch.

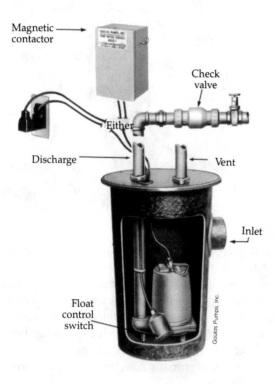

Magnetic contactor

Check valve

Either

Discharge

Vent

Inlet

Float control switch

Goulds Pumps, Inc.

Fig. 21-9. Typical sewer pump setup.

The two biggest problems with laundry pumps are the check valve and the strainers. If the check valve will not open, the pump cannot pump. If the check valve cannot close, the pump will be forced to continually pump the same water over and over. If the strainer or the impellers are jammed with sediment, lint, or other obstructions, the pump will not work properly.

Final words

Before working on any equipment, refer to the owner's instructions. Be careful around the electrical aspects of the equipment. There are numerous types of plumbing fixtures and equipment. The information in this book is based on average conditions with average equipment. Your particular situation or equipment may call for different techniques. The suggestions given in this book should help you solve problems, but if you continue to have trouble, refer to the owner's manual or call the dealer. If all else fails, call a professional to correct the problem.

By now, you are well informed about basic professional plumbing procedures. With this book and a little common sense, you should be able to tackle most plumbing jobs. Know your limitations and don't attempt any job for which you are not competent. I wish you the best of luck with all of your plumbing endeavors.

Index

cutting out new plumbing, 54-
56

D

defective valves, 236
directional fitting changes
  table, 47
dishwashers, 147-149
  connecting the water supply,
    148-149
  symbols, 38
drain, waste, and vent (DWV)
  systems (see also drain
  complaints), 24-27, 211-212,
  230-231
  brass pipes, 24, 26
  building drain and sewer, 48
  cast-iron pipe, 24, 26
  copper pipe, 24, 26
  cutting the pipe, 211-212
  dry vents, 83, 86-87
  galvanized steel pipes, 24,26
  horizontal fixture drains, 48
  installing the DWV system,
    80-85
  lead closet bends, 212
  lead traps, 212
  leaving old pipe in place,
    211
  material take-off table, fit-
    tings, 59
  planning the system, 82
  putting the system together,
    87-89, 91-95
  running pipe, 82-83
  schedule 40 plastic pipe, 25-
    26
  size table, 47
  sizing the DWV system, 46-
    50
  test balls and caps, 230-231
  tools, 80, 211
  vents, 48,49
  wet vents, 83
drain complaints, 221-232
  basket strainers, 225
  bathtub drains and traps,
    223-224
  frozen drains, 230
  overloaded drains, 231

municipal failures, 231-232
septic tank backups, 231
shower drains, 224
sink drains, 224-225
toilet drains, 225-228
vent problems, 221-222
drainage fixture valve table, 46

E

electric water heaters, 242-244
  heating elements, 242
  reset buttons, 243-244
  temperature adjustments,
    244
  thermostats, 243
electrical failures, 234

F

faucets, 208-210, 248
fixtures, 27, 45, 47
  capacity table, 44
flame striker, 23
flaring tools, 7
floor drain symbol, 40
flush holes, 225-226
flush valves, 226
flux, application to pipes, 22
frost-free sillcocks, 245
frozen drains and vents, 222,
  230

G

galvanized pipes and fittings,
  17, 19, 24-26, 192-193, 197-
  198
  rubber couplings, 192
  threaded adapters, 192-193
garbage disposers, 145-147,
  216-217
gate valve, 98
grade levels, 10-11

H

hacksaws, 3-4, 3
hammers, 13
hand drills, 15
hand tools, 4-13
  adjustable wrenches, 8
  basin wrenches, 8
  chain wrenches, 7
  cold chisels, 12

conventional levels, 11
convertible screwdrivers, 12
cutters, 4-6
flaring tools, 7
grade levels, 10-11
hacksaws, 3-4
hammers, 13
hex wrenches, 8
line level, 11-12
metal-cutting snips, 9
miniature roller cutters, 5-6
needle-nose pliers, 9
phillips screwdrivers, 12
pipe cutters, 6
pipe wrenches, 7
roller-type cutters, 4-6
safety, 15
shovels, 13
square-blade screwdrivers,
  12
strap wrenches, 8
tongue and groove pliers, 9
torches, 13
torpedo levels, 11
tubing benders, 10
universal saws, 13
utility pliers, 9
wood chisels, 12
hard water, 161
hex wrenches, 8
hose bibs, 41, 169

I

ice makers, 149
inspection and testing, 164-170
  bathtubs and showers, 168
  hose bibs, 169
  kitchen plumbing, 168-169
  laundry hook-ups, 169
  lavatories, 167-168
  miscellaneous plumbing, 169
  toilets, 164-166
  water heaters, 169
  water pumps, 169
installation procedures, 127-
  149
  bathtubs without shower
    surrounds, 127-128
  conditioners and pumps,
    152-163

Other Bestsellers of Related Interest

WELLS AND SEPTIC SYSTEMS—2nd Edition
—Max and Charlotte Alth,
Revised by S. Blackwell Duncan

Revised to conform to current codes and requirements, this expanded edition examines well and septic tank installation, water storage and distribution, water treatment, ecological considerations, and septic systems for problem building sites. It covers everything from obtaining necessary permits and locating water and percolation test sites to selecting and preparing a septic system. 272 pages, 156 illustrations. Book No. 3824, $16.95 paperback, $25.95 hardcover

THE HOMEOWNER'S GUIDE TO DRAINAGE CONTROL & RETAINING WALLS
—Jonathan Erickson

New homebuilders and veteran homeowners alike will appreciate the practical information this sourcebook provides on how to deal with drainage problems. This guide includes a comprehensive overview of retaining walls and drainage systems and instructions for building retaining walls made of wood, stone, concrete block, or brick. Illustrations are used throughout to clarify the concepts and principles explained. 160 pages, 124 illustrations. Book No. 3153, $14.95 paperback only

HOME HEATING AND AIR CONDITIONING SYSTEMS—James L. Kittle

Spare yourself the aggravation of trying to locate a repairman when your furnace or air conditioning system breaks down—do your own professional-quality maintenance and repair! With the comprehensive instruction and guidance included here, you can install, repair, or replace just about any type of home heating or air conditioning system—oil, gas, hot water, or forced air heating systems, and central air systems, heat pumps, and more. 272 pages, 198 illustrations. Book No. 3257, $15.95 paperback, $24.95 hardcover

CERAMIC TILE SETTING—John P. Bridge,
Photography by Robert A. Bedient

Discover how easy it can be to install your own ceramic tile floors, walls, and counters for a fraction of what you'd spend to hire a pro. From initial layout to floating and leveling, this easy-to-use guide contains all the information you need to start and finish a professional-looking project. Projects are arranged in order of difficulty and include step-by-step instructions. 244 pages, 165 illustrations. Book No. 4053, $14.95 paperback, $24.95 hardcover

ROOFING THE RIGHT WAY—2nd Edition
—Steven Bolt

Why pay a contractor thousands of dollars to put a new roof on your home when you can do it yourself? Roofing isn't as difficult or as costly as you might think. Following the guidelines presented here, you can install a watertight roof that will add to the beauty and value of your home for years. You'll find in-depth details on ever aspect of roofing—from choosing the proper tools and materials to step-by-step application techniques for nearly any type of roof. 240 pages, 277 illustrations. Book No. 3387, $14.95 paperback only

PROFESSIONAL PLUMBING TECHNIQUES: Illustrated and Simplified—Arthur J. Smith
". . . useful for the experienced handyperson or the homeowner who wishes to verify an estimate of work needed." **—Booklist**

This plumber's companion includes literally everything from changing a washer to installing new fixtures: installing water heaters, water softeners, dishwashers, gas stoves, gas dryers, grease traps, clean outs, and more. It includes helpful piping diagrams, tables, and charts. 294 pages, 222 illustrations. Book No. 1763, $14.95 paperback only

MICROWAVE OVEN REPAIR—2nd Edition
—Homer L. Davidson

Save on repair bills and reduce your microwave's downtime with the fix-it advice in this manual. You get complete coverage of the latest microwave oven models and their features, detailed maintenance and troubleshooting guidance, test procedures used by major appliance manufacturers, and specific solutions to more than 200 common microwave oven malfunctions. 384 pages, 366 illustrations. Book No. 3457, $29.95 hardcover only

HOME WIRING FROM START TO FINISH
—Robert W. Wood

Safely and successfully wire an entire residence with this do-it-yourself manual. Two-color illustrations and photographs make it easy to identify components and follow Wood's step-by-step directions. For personal wiring jobs in your own home, this guide will make sure you pass inspection every time. 272 pages, 421 illustrations. Book No. 3262, $26.95 hardcover only

Prices Subject to Change Without Notice.

Look for These and Other TAB Books at Your Local Bookstore

To Order Call Toll Free 1-800-822-8158
(24-hour telephone service available.)

or write to TAB Books, Blue Ridge Summit, PA 17294-0840.

Title	Product No.	Quantity	Price

☐ Check or money order made payable to TAB Books

Charge my ☐ VISA ☐ MasterCard ☐ American Express

Acct. No. _____ Exp. _____

Signature: _____

Name: _____

Address: _____

City: _____

State: _____ Zip: _____

Subtotal $ _____

Postage and Handling
($3.00 in U.S., $5.00 outside U.S.) $ _____

Add applicable state and local
sales tax $ _____

TOTAL $ _____

TAB Books catalog free with purchase; otherwise send $1.00 in check or money order and receive $1.00 credit on your next purchase.

Orders outside U.S. must pay with international money order in U.S. dollars drawn on a U.S. bank.

TAB Guarantee: If for any reason you are not satisfied with the book(s) you order, simply return it (them) within 15 days and receive a full refund. BC